AF615222

BEYOND THE BEAKER

TO DAVE,

THE BEST OPERATIONS GUY I'VE EVER WORKED WITH.

ALL MY BEST,

Paul Patton

6/4/2009

BEYOND THE BEAKER

HOW TO ACHIEVE SUCCESSFUL MARKET ADOPTION FOR EMERGING TECHNOLOGIES

Paul Patterson

Kohritsu Press

Published in the Unites States of America by Kohritsu Press LLC
Printed in the United States of America

Library of Congress Control Number: 2009901683

ISBN 13: 978-0-578-01111-0

The paper and ink used in this publication meet the requirements of the American National Standard for Permanence of Paper for Publications and Documents in Libraries and Archives Z39.48-1992

Acknowledgements

First and most appreciative acknowledgement goes to the authors of the previously published content referenced throughout this book, and to their gracious support for this publication. It also goes to the hard work employed by all the innovators of emerging technologies referenced herein.

A special graditude goes to Ms. Wai Yee Chan for her persistance and patience with the author while editing content. Her insightful business acumen resulted in numerous restructurings of content and is credited for pushing to completion.

Other people who have provided assistance have been fellow professionals for whom without their support, this book would not be here.

Yet with considerable debt to all, it goes without saying that none of them should be held responsible in any way for my inability to fully benefit from their assistance and make this a better book than it is.

In graditude to all
Paul Patterson

Contents

BEYOND THE BEAKER

Preface

"Thinking is easy, acting is difficult, and to put one's thoughts into action is the most difficult thing in the world." – Johann Wolfgang Von Goethe

Identifying new opportunities, building them into a viable businesses or launching them into a sustainable company is tough. The process is filled with hurtles and pit falls, which must be overcome to achieve success. This book focuses on the process of moving emerging technologies from the laboratory to the market. Fundamental business processes are often over looked and result in business failure, stagnation, late adoption, or launched into a commodity market. This book describes the process of performing strategic business planning for emerging technologies. It is not about introducing products into markets. It is about demonstrating how a process can be used, exampled by case studies, to integrate business practices such as; Corporate Portfolio Strategy, Product Portfolio Management, Brand Portfolio Management, Business Designs, the fundamentals of Non-Disclosures, Joint Development and Joint Venture Agreements, and numerous other business practices into a process.

Case studies are commonly written as if everything was planned, great results occurred and summarized with three take-aways. This process can set unrealistic expectations for the reader. The true, *Real Life* events which occur and are more often the critical events resulting in success are frequently not documented. A couple colleagues have requested that this book include details of events such as, it being in Beijing, its 4 am and your phone rings. The person on the line is upset, screaming vulgarities because someone in the collaboration violated international trade laws. It's your job to repair the situation. Or, a tier one company burning a bridge with an exclusive material supplier in Japan and was told, "Leave, get your material from some other company!" It's your job to repair the relationship. Or, due to poor coordination of activities, a multi hundred million dollar collaboration is canceled because another department simultaneously and independently threatened lawsuit against the collaborator for IP infringement, resulting in several months of meetings and relationship management. Or, a production line in China is duplicated on the other side of a wall for which the foreign company is not allowed to see; subsequently resulting in significant loss of market share. They wanted this book to include experiences around scheduling conference calls from 2 am to 4 am because that's the only time available for collaborators around the globe to meet. This is the real work behind the scenes which are unplanned and must be managed. The work nobody writes about, but discuss over dinner meetings. These should be documented to provide realist expectations to readers. This book attempts to demonstrate through case studies, examples of these situations and how they were managed. The content is structured to coincide with activities related to the management of emerging technologies.

Emerging technologies typically start in a laboratory. This is where an inventor achieves their "Eureka!" The laboratory can be an elaborate, fully funded corporate laboratory, a university laboratory, or even an inventor's garage. Some believe emerging technologies are limited to electronic technologies when in fact they are found in virtually every market sector. For example, in the cosmetics sector, the chemistries derived to enhance the interaction between shampoos and one's hair can be very complex. The discovery that Fullerenes are more than 125 times more effective than vitamin C as molar ratio is beginning to impact the cosmetics market to enhance anti-aging and antioxidant effects. Often, when the inventor achieves their Eureka, they obtain a vision of running a large multi national corporation, completely unaware of the complexities to achieve such a goal. As will be demonstrated, there is far more to managing and building an enterprise than achieving the "Eureka!"

Examples of the knowledge base required beyond the eureka are: intellectual property management and strategy, logistics management and strategy, identifying the required collaborative technologies and companies, determining value proposition, educational and cultural aspects, identifying and managing stakeholders, contract negotiations and many others discussed throughout this book. Whether the technology is managed from within the corporate environment or launched as a startup, there are numerous factors, which must be planned and executed.

Assessing the technology from a fundamental perspective of maturity, application, and intellectual property position is the first step. The process identified in chapter 11 exposes the reader, exampled by the Poly-OLED case study, providing detailed accounts of issues and solutions. The full prospectus of the encompassing process is similar to a puzzle. Identifying the technology and its application is only a small piece of the puzzle. Case studies mentioned are comprised of historical and current emerging technologies. Examples of historical are the 3M Post-It® note and the internet. Examples of current emerging technologies are Cambrios and a detailed case study on Cambridge Display Technologies (CDT), a leading developer of technologies based on polymer light emitting diodes (Poly-OLEDs).

As one who has been involved with numerous emerging technologies around the globe, over time, inconsistencies have become apparent. Both successful and non-successful practices are exhibited. Sometimes, one can be shocked when exposed to some strategies or how they are executed. Other times one can be impressed with professional execution and strategic direction.

Inventors and investors alike, drive forward with expectations of achieving success. Their goals start out the same; a desire to achieve market adoption. Market adoption for an investor could mean a respectable return on investment through an exit strategy: acquisition or IPO. While a founding entrepreneur may view successful adoption as attaining market leader position. Success rates are not 100%. Success rates vary across different investor types and technology sectors. However, many startups fail to simply get off start, and many fail due to numerous other factors. Listed here are common factors for failure.

- Breakdown of investor and inventor relationship
- Market dynamics
- Funding problems
- "Me too" technology
- No channel to market
- Wrong target market segment
- Strategy (Poorly designed and/or executed)
- Team experience (Inventor & Investor)
- Loss of strategic control
- Technology fails to achieved expectations
- Poorly managed intellectual property
- And many more that will be discussed

This book was written in an attempt to alleviate the majority of these failure factors. It is a comprehensive text identifying challenges, key success factors and insight on how to manage them. It is intended to be a compilation of knowledge with the goal of increasing the knowledge-base on best practices for managing emerging technologies, allowing the reader to leverage and expand upon the content. The purpose here is to enhance the stakeholder's expertise with a comprehensive commercial awareness, to help them realize their aspirations and make their innovations a reality.

This book provides insight to the numerous professional talents required to bring an emerging technology to market adoption. It is important to understand that the knowledge contained within, will not make one an expert in the field for any one of these professions. The intention of the discussed insight is to bring an understanding of the significance of hiring the right talent, one who could bring the extensive experience and knowledge to the team.

Chapter 1

Introduction

Success is going from failure to failure without a loss of enthusiasm.
– Winston Churchill

In 1893, Leo Baekeland, a young Belgian immigrant to the United States had developed a new type of photographic paper. This paper was prepared with silver chloride emulsion, which eliminated the washing and heating steps of image development. Baekeland's new paper could be exposed by artificial light enabling amateur photographers to develop their photographs at home or in one of the new processing laboratories. Desperate and on the verge of bankruptcy, Baekeland decided to sell his new photographic paper to Eastman Kodak.

While taking the train to meet with Eastman Kodak, Baekeland decided he would ask for $50,000 and if forced to compromise he would settle for no less than $25,000. Eastman Kodak was so impressed with Baekeland's photographic paper, he offered Baekeland $750,000.

Sounds like a simple process to follow: create an emerging technology; call up the industry leader; offer to disclose the technology; the industry leader says, "Sure come on out and show us your technology", then offers 15 times the inventor's perceived value - no Non-Disclosure Agreement (NDA), no Joint Development Agreement (JDA); the industry leader clearly understands the value of the technology, and you have easy access to the decision makers.

However, this is the 21st century. Non-Disclosure Agreements are required before anything confidential is discussed. Legal counsel is brought in early. Access to the decision makers is not only difficult, sometimes impossible. Joint development efforts are commonly required when managing emerging technologies. Perceptions of technological limitations act as hurdles and stop gates. Value justifications are only a small part of the negotiation process. There is a common misconception that managing an emerging technology is easy.

There are cases where a technology has been rapidly adopted but these are rare. As demonstrated in the proceeding pages, a more realistic and common adoption rate takes significantly longer and success is dependant upon numerous factors.

Adoption Rates of Emerging Technologies

Emerging technologies are technologies which commonly originate in a laboratory. Most of them begin at the materials level and require collaborative technologies prior to adoption into a product. These laboratories can be located at a university, a corporation, or even someone's garage.

The adoption of emerging technologies can range from a few years to several decades. For example, the graph below depicts the technology adoption rates for several technologies in the USA communications sector.

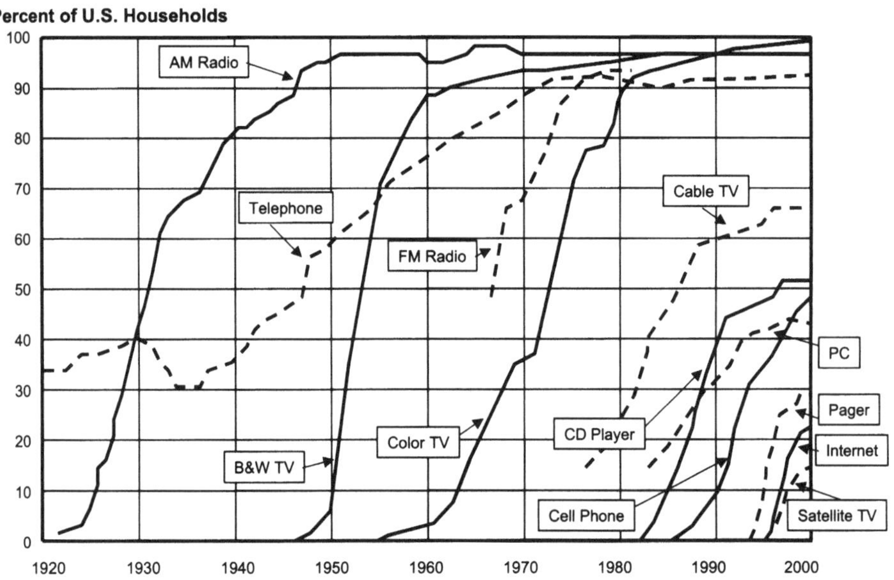

EXHIBIT 1.1 Adoption Rates of Various Communication Technologies (USA)

Source: The Wall Street Journal Classroom edition, 1998

This 80-year historical chart shows the different times required for several technologies to progress from the conceptual stage to the high adoption stage. It demonstrates that some technologies were adopted faster than others. Adoption rates are affected by numerous factors. The telephone took 60 years to reach the high adoption rate, mostly because it required a large infrastructure investment before consumers were able to use it. On the other hand, the Internet was adopted relatively quickly in ten years because it was able to take advantage of the existing telecommunications infrastructure and computer literacy of the population.

The rate of adoption is influenced by both macro factors that one cannot control, and variables that one can manage. Macro factors normally beyond one's control are political, economical, legal, social and technological. However, the variables that are mostly organizational and commercial can be managed to increase success rate and speed of adoption of the emerging technology.

Innovation

There have been numerous books and articles published about innovation. Some publications classify innovation according to its relative social impacts and sectors. Innovation, simply put, is a solution to a problem. Famous ones are the wheel or the lever. A more modern example is the solid state transistor. When Seagulls pick up a crustacean, fly up a few meters and drop it, to crack it open, solves the problem of not having the ability to crack it open with their beaks: this is also innovation. The 'act of conceptualizing the use' of twisted cholesteric liquid crystal to create a bi-stable reflective display technology (Reflex™) is innovation. The resulting display that was created became an emerging technology. The further conceptualized use of Reflex™ to cover the cases of consumer electronics, enabling consumers the ability to digitally change the color of their device, is further innovation. This also provides an example of an originally unforeseen application for an emerging technology. With the numerous publications on innovation, there is not much published with a tangible take-away.

What steps does an organization need to perform to create an innovative environment? The Epson Portland Inc. (EPI) case study will be exampled in an attempt to answer this question.

EPI, a wholly owned subsidiary of Seiko Epson Corporation (Japan), was founded in 1985. In January of 2001, EPI's primary business units were ink cartridge (ICBU) and printer manufacturing. They competed in the global market for high volume manufacturing with 1100 employees. Also in January of 2001, the president transferred management of the ICBU to the VP (Dave Graham). At that time, the total defect rate for in-process production was not being calculated. Dave pushed for metrics, within 60 days it was determined that production was experiencing more than 11,000 DPM (Defect Per Million). This number was calculated only from the automated lines and excluded plastics and packaging. EPI was plagued with serious customer complaints from Epson's logistics centers. These claims included such issues as no ink, ink leaking, and other serious functional defects. Due to the seriousness of the claims, employees were in a constant fire fighting mode making it difficult to make progress on process improvement. Operations were experiencing a 35% turnover rate. Top executives in Japan were concerned about the possible harm to the Epson brand and were relatively sure the operation would have to be closed.

Plagued with manufacturing issues coinciding with changes in tax laws, it was announced on April 30, 2001, that EPI's printer assembly business would have to be closed and moved to China. Within 6 months, EPI's head count dropped from 1100 to 250. It took until October to disposition the associated assets. Several reorganizations were implemented throughout 2001. Dave Graham was promoted to EVP. Headquarters was not pleased with the performance of the ICBU and were looking to transfer its operations as well. In an attempt to save EPI, Dave made several trips to Japan making promises to achieve aggressive goals to increase quality, reduce cost and improve delivery.

To avoid being closed, EPI had to change. Its culture was characterized by hiding mistakes, poor communication, lack of cooperation, poor discipline, limited teamwork, placing blame on one another and resistance to change. In October and November, the management within the ICBU operation was replaced. A new director was hired from outside the organization and a high-level engineering management experienced assignee from Japan was transferred to EPI. Dave worked with the remaining EPI management to develop a new vision, mission and value system. Aggressive plans were set in place with concrete targets and milestones.

Communications with employees were increased since the April announcement. Monthly production results and the quarterly action plan were reviewed with ALL employees. Every meeting included a Q&A session. Even financial data were shared. An open door policy was created so that any employee could talk to the EVP directly if they felt it necessary. Dave made it clear that survival of EPI is dependant upon every employee. "It's either turn the company around and compete with China and Mexico or dust off our resumes (CV)", Dave concluded.

An employee innovation program was initiated. Leveraging quality processes from Toyota and other successful manufacturing organizations, EPI implemented a Kaizen program designed to engage the entire organization, at every level. The program rewarded employees for suggestions of improvement. Employees could post suggestions on the bulletin board, on line, or fill out a form and submit. Each suggestion was calculated to determine the financial impact to the organization. Points are awarded for varying levels impact. Financial awards were granted for achieving three different levels. Rewards were extended to temporary employees as well. These rewards were announced at the all-employee meetings.

Implementing an employee participating innovation system (Kaizen), open communications and sharing of financial results built a trust with management. Every employee knew their value (contribution) to the organization. The organization began to turn around. On May 2, 2002, Epson's Chief Production Officer (CPO) visited EPI to review the metrics and tour the factory. Upon completion of his review, the CPO told the employees of EPI that, "They had performed a miracle". His advice and encouragement inspired the employees to achieve even higher goals. By the fall of 2002, EPI was the first production facility to attain the corporate goal presented to all global operations of achieving a DPM of 300, for total facility operations.

The employees of EPI achieved the goal and continue to set more. Additional production lines have been brought to EPI. This is a global economy; EPI is proof that through innovation, team work, committed management and open communications, any organization can compete. As Kaizen principles are characterized by continuous improvement, EPI's Kaizen process is also continuously evolving. They have been kind enough to provide a representative sample of their process, which can be found in Appendix C.

As demonstrated by EPI, Innovation occurs at ALL levels of an organization. Some believe that innovation only occurs in product development. This belief is simply due to their point of reference. In 2003, Jean Cunningham and Orest Fiume published *Real Numbers*, where they demonstrated how kaizen principles applied to accounting can dramatically enhance organizational performance and reduce costs. Solutions to problems (innovation) can occur throughout an organization, if the culture and processes are in place.

An *Emerging Technology* is simply a bi-product of innovation, which in turn can inspire additional innovation. The solid state transistor has been noted as one of the most significant developments of our time. It replaced the old vacuum tube designs and made transistor radios possible. The transistor was, and continues to be, an emerging technology. It evolved from innovation of materials and processes. However, utilizing the transistor to solve or simplify problems in varying applications is continued innovation. This book focuses on effective management of emerging technologies. Innovation and organizational implementation thereof is a different subject.

Emerging Technology Market Roadmap – Two Different Perspectives

When an emerging technology discovery occurs, its founders quickly identify applications for use. The field of innovation is constantly in flux, several added every day, as some fall to the side. Managers and investors of emerging technologies are frequently searching for the "Killer Application", while corporations are searching for the "Killer Technology" enhancing strategic advantage. Both are primarily looking for "Differentiation" which provides market advantage. Objectives and processes are different for each. Yet both are centered on emerging technologies. While attempting to solidify adoption, the emerging technology refines itself and performs processes such as Customer Identification. However, the corporate approach is to seek out emerging technologies that compliment their core competencies. Exhibit 1.2 identifies the roadmap for emerging technologies, from the market push of the emerging technology to the corporate or market pull.

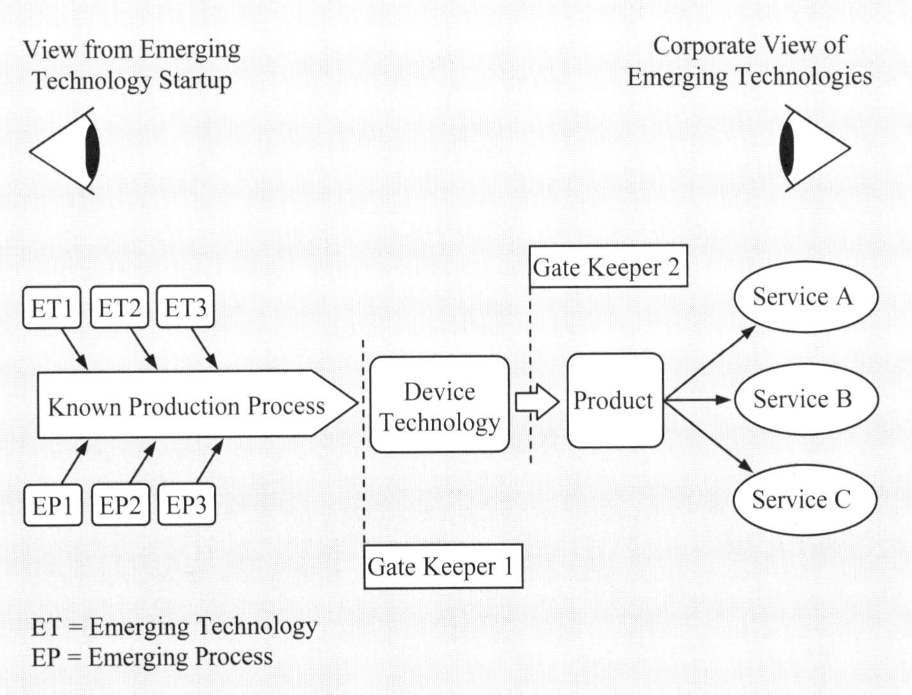

EXHIBIT 1.2 Emerging Technology Market Roadmap

There are many emerging technologies vying for position into known processes. Each promising improved performance and/or reduced costs. Emerging technologies often focus exclusively on identifying the potential buyer who can use the technology in their process. Their efforts commonly stop there. When in reality the true value is delivered further down the supply chain. Likewise, corporations often rely exclusively on their direct suppliers to be the provider of emerging technologies. Reliance on the supplier to be the provider of the emerging technology empowers the supplier with a gate keeper function. These two approaches generate a gap; where as corporations are not aware of lower level technologies that can enhance their product offerings and emerging technologies are not aware of potential market applications beyond the product level. Bridging this gap can provide access to "killer" applications for the emerging technology and provide enhanced user experiences for service level providers.

The market roadmap can be comprised of a varying number of gate keepers. In the cases where organizations have successfully managed vertical integration, the numbers of gate keepers are minimized. Samsung is both a display manufacturer and a mobile phone manufacturer. Samsung also performs extensive research on innovation, developing emerging technologies and processes for internal use. This vertical integration reduces the number of gate keepers. Adoption is left to the service providers and the consumer.

This emerging technology roadmap applies to every technology sector. It is the consumer or government agency who makes the final decision of adoption. This book will identify through a series of case studies best practices from the market and from literature to successfully manage emerging technologies, bridging the gap between these two different perspectives.

Book Format

Frequently termed "Managing Innovation", in-depth knowledge and perseverance are the essences in managing an emerging technology. To achieve industry adoption, it requires comprehensive, multidisciplinary understanding, a good team, ability to make tough decisions and most importantly, tenacity.

There are already an extensive number of books written on the technology development process, designed to cover the different phases and requirements for a technology to achieve production. This book comes from a different angle, focusing on the business aspects that are critical and oftentimes more important to the success of the technology. This book is comprised of a selection of previously published content, compiling it within a structured framework integrated with case studies so that one can identify the value brought and what role they play with emerging technologies. It highlights by example the types of pitfalls and market dynamics one must be prepared to manage.

The target audiences for this book are innovators and managers of emerging technologies. Startups based upon an emerging technology are typically founded by innovators, who are individuals with extensive technical knowledge and experience. The purpose here is to enhance the innovators' technical expertise with a comprehensive commercial awareness, to help them realize their aspirations and make their innovations a reality.

This book is comprised of three sections:

Section One (Issues relating to managing an Emerging Technology Startup): Chapters 1-8 will discuss the origins of Eureka and the critical aspects to be managed from technology development, strategic marketing, business development, legal considerations, financing through organizational management.

Section Two (Corporate Perspective of Emerging Technologies): Chapter 9 will provide perspectives from potential investors such as business angels, venture capitalist and corporate ventures. Chapter 10 will provide a framework for managing emerging technologies from a corporate perspective.

Section Three: Chapter 11 is a case study on Poly-OLED, utilized by example, highlighting the common pitfalls and challenges that emerging technologies can undergo. Chapter 12 summarizes the findings disclosed throughout the book. Chapter 13 is an open discussion about a few currently emerging technologies and their differentiation.

The secondary audiences for this book are business angles, corporation ventures, venture capitalists and corporate spinouts. Understanding the different challenges that an emerging technology startup undergoes during different technological phases will help them to work alongside or manage the startups better.

Chapter 2

Emerging Technologies

An inventor is a person who makes an ingenious arrangement of wheels, levers and springs, and believes its civilization.
- Ambrose Bierce

Hype Cycle

Emerging technologies typically start with a single individual, who recognizes a potential opportunity that most individuals would overlook. On February 9th, 1989, at Cambridge University in the UK, Dr. Jeremy Burroughes was studying the Franz Keldysh effect on conjugated polymers under the supervision of Dr. Sir Richard Friend. While setting up the experiment in the Cavendish Laboratory, which involved wiring 100 micron gold wires to a cryostat, he noticed another experiment on a similar device about five feet away. Glancing over at the other device, Dr. Burroughes noticed it was emitting green light. Initially he thought it was an accidental reflection from the green phosphor in the computer monitor. Upon moving closer to the screen, he realized the light moved across the screen, quickly dispelling the idea that the computer monitor was the light source. As Dr. Burroughes turned the voltage source off, on the other experiment, the light went away. Believing this to be an opportunity, he copied the contact arrangement of this experiment and connected the Franz Keldysh device in exactly the same way. He slowly increased the voltage on his device until light was emitted. At that moment, he discovered a method to emit light from plastic. This was the birth of Poly-OLED (Polymer Light Emitting Diode). This first device emitted light for three days.

Immediately, Dr. Burroughes could envision the potential value of this technology within numerous future applications. He and his colleagues first approached the university but they quickly discovered the university had no means of filing patents. So they located a patent agent and filed the patent for themselves.

After the discovery, it still required someone to nurture the concept and transform it into a viable technology. Dr. Jeremy Burroughes' experiment occurred in 1989. The first working displays comprised of 3X5 pixel arrays were developed in 1991.

To achieve industry recognition the technology required the establishment of a corporation and investor funding. With seed capital funding, it was rolled into a company named Cambridge Display Technologies (CDT) in 1992. It was not until 2002 that this technology became increasingly adopted in consumer products. Philips launched one of the first consumer products containing OLED technology in their innovative electric shaver with an electronic display, which was featured in the James Bond movie "Die Another Day". More than 15 years after the discovery, cell phones and PDA's became increasingly available with Poly-OLED technology.

Envisioning the future, other applications for Poly-OLED technology were demonstrated in movies. This James Bond movie covered a car with Poly-OLED for camouflage; it displayed the background of the surroundings on the surface of the vehicle, rendering it invisible to the viewer. In the movie "Minority Report" an electronic newspaper was digitally updated to report a fugitive (Tom Cruise) on the loose. As the technology matures, time and innovation will tell how and where it will be applied.

The Poly-OLED discovery was never planned; it emerged as an unexpected result from research. As will be discussed in the next chapter, discovery occurs and is never planned for. The Poly-OLED technology underwent the typical hype cycle process for emerging technologies. Since 1995, Gartner has used Hype Cycles to describe the typical progression of an emerging technology, from over enthusiasm through a period of disillusionment to an eventual understanding of the technology's relevance and role in a market or domain According to Gartner, "Hype Cycles also show how and when technologies move beyond the hype, offer practical benefits and become widely accepted." The Gartner Hype Cycle is depicted in Exhibit 2.1.

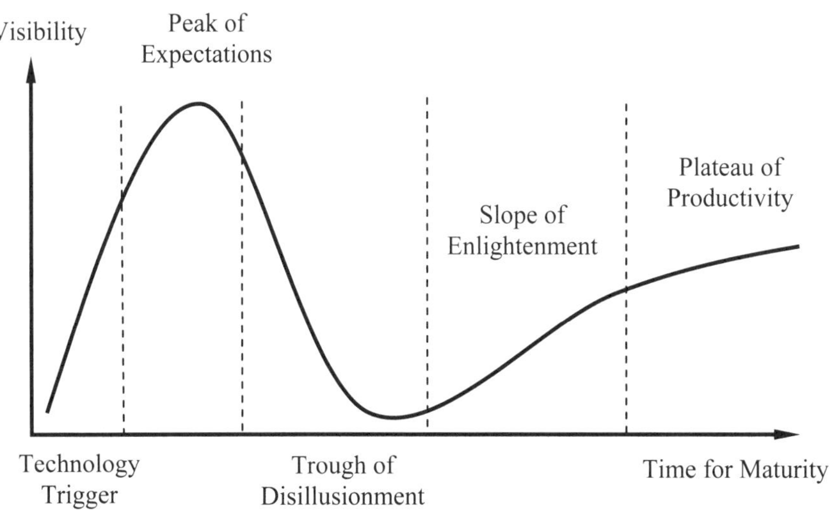

EXHIBIT 2.1 Gartner Hype Cycle

Source: Gartner, Inc., "Understanding Gartner's Hype Cycles, 2008", by Jackie Fenn, June 27, 2008
Reproduced with permission

The road from the laboratory to the market is an enduring one filled with numerous obstacles. As a technology progresses down this road, it will experience several ups and downs. Gartner Inc. depicts the typical life cycle of emerging technologies in their Gartner Hype Cycle. The Gartner hype cycle is described as a five-phase sequence:

1. Technology Trigger
   "The first phase of a Hype Cycle is the "technology trigger" or breakthrough, product launch or other event that generates significant press and interest."

2. Peak of Inflated Expectations
   "In the next phase, a frenzy of publicity typically generates over-enthusiasm and unrealistic expectations. There may be some successful applications of a technology, but there are typically more failures."

3. Trough of Disillusionment
   "Technologies enter the "trough of disillusionment" because they fail to meet expectations and quickly become unfashionable. Consequently, the press usually abandons the topic and the technology."

4. Slope of Enlightenment
   "Although the press may have stopped covering the technology, some businesses continue through the "slope of enlightenment" and experiment to understand the benefits and practical application of the technology."

5. Plateau of Productivity
   "A technology reaches the "plateau of productivity" as the benefits of it become widely demonstrated and accepted. The technology becomes increasingly stable and evolves in second and third generations. The final height of the plateau varies according to whether the technology is broadly applicable or benefits only a niche market."

Applying the Gartner Hype Cycle to the Poly-OLED discovery, the *Technology Trigger* occurred in 1989 and it generated significant amounts of interest. As the technology climbed towards the *Peak of Expectation*s, CDT decided to build a clean-room fabrication facility in Godmanchester UK for small-volume production of display panels. However, around December 2002, the display manufacturers began to realize that the material suppliers still required additional time on research. The technology was premature for the market and fell into the *Trough of Disillusionment.* These companies stopped announcing successes at conferences and decided to focus on solving the remaining issues surrounding the technology. This silence from the industry resulted in the perception that the industry had moved away from the technology. Investor interest and magazine publishers followed this perceived direction. With investor funding for CDT becoming increasingly difficult to obtain, stakeholder management became the highest priority for CDT. When a technology finds itself in the *Trough of Disillusionment*, the effort required to realize the true value can take a significant amount of time and dedication; coupled with patience. Only by making difficult decisions was CDT able to survive. CDT modified its business model by abandoning small volume production at the Godmanchester facility, refocused R&D and laid-off 25% of its resources. In a collaborative effort, the material suppliers and display manufacturers began to focus on discovering the true limits and capabilities of the materials; this migrated the technology onto the *Slope of Enlightenmen*t. The subsequent eighteen months were filled with display manufacturers demonstrating proof-of-technology in display sizes from 2 to 17 inches. In May of 2004, Epson announced a breakthrough by demonstrating the world's first 40-inch inkjet printed Poly-OLED display.

Throughout this process, stakeholders must be managed. Stakeholders can include inventors, investors, material suppliers, and everyone involved in bringing the technology to the consumer. Managing the stakeholders can be the most tasking part of the process required to bring an emerging technology to market. Underestimating the significance of stakeholder management can be devastating. Several things can - and will - go wrong. Suppliers can quit producing or even collaborate on blocking strategies, investors can lose confidence, and intellectual property can get out of control creating barriers to market.

As an example, let's use this graph to discuss the Internet and all its promises. In the beginning, the Internet was only used for data transfers between military contractors, government organizations and educational institutions. The emergence of application software and search engines launched it into a powerful communications tool. Eventually, the Internet was hyped to be the only place to do business. Traditional over-the counter businesses were predicted to fail. Subsequently, the e-commerce bubble burst. Corporations and investors came to the realization that the Internet is merely an additional channel for conducting business. The Internet could not replace sound business practices.

The Origins of Eureka!

Where or how does the Eureka originate? What are its foundations? Universities and corporations have attempted to define innovation with a goal of exploiting its value. An ever-increasing number of Universities offer an accredited degree in Innovation. There is a common misconception that ALL innovation starts with a clear idea or eureka. This is a nice picture, which most stakeholders like very well. But this is a poor reflection of reality. Innovation is far more complex. For example, it is much more probable that during the development stage of a product a new application is produced as a by-product, which was neither planned, nor existed as an idea in the beginning.

Pauchard's Innovation Engine

Dr. Marc Pauchard of ILFord Imaging located in Fribourg, Switzerland, created a diagram to demonstrate how emerging technologies require both internal and external relationships to become products, Exhibit 2.2. The diagram provides a pictorial view from a technology perspective of the relationships required throughout the process.

The diagram starts with an Idea or eureka, a product or technology concept, which originates from either an internal or external source. The diagram has corporate core competencies located at the middle of the main drive gear. This indicates development efforts should focus around the corporate core competencies. The neighboring gears are comprised of shared technologies between external competencies and internal competencies. This shared technology is commonly protected with either an NDA or JDA, maintaining the technology within the control of the corporation.

As the Innovation Engine turns, basic concepts start in the infancy stage, where fundamental and advanced discussions and experiments occur to develop the technology to the next phase. Not all competencies are within the organization and may require utilizing external sources. These can be industry professionals or university experts. As the technology becomes mature, processing or manufacturing requirements will need solutions.

Again, the corporation may need to utilize external resources or collaborators to assist in the development process for manufacturability. Eventually, the technology reaches maturity, becomes manufacturable, can be applied to applications, and submitted for business development.

Innovation can happen at any level. It can be a "Technology Push" or a "Market Pull" that sets off the beginning of a success story. Pauchard's message is that one never knows when and where innovation is going to happen. The only thing one can do is create an environment for innovation to happen and be ready on all levels to accelerate the engine up to speed, once an idea for a "Killer Application" surfaces. There is no value in a great business idea, if it is not technically feasible and vice versa.

The highest probability for innovation to happen in a way that is beneficial for the commercial success of a company is to stay close to the field of competence: there, your people know best to distinguish between high flights of fancy and real opportunities! Even more: on their field of competence they have developed good "intuition". "Intuition" is much more critical to innovation than "planning" will ever be.
This approach appears to be slightly out of alignment with the teachings of Cooper et al. in that corporate resources should be in alignment with corporate strategy, whereby new technologies are provided as directives derived from strategy. Not all technologies emerge from such a direction. When a Eureka is realized or identified, it must be integrated into the corporate strategy and undergo a process similar to the diagram described by Pauchard. Attempting to integrate a Eureka before realization would simply be attempting to predict innovation, a task proven to be difficult at best. What the diagram clearly identifies is that success (defined as market adoption) requires far more resources and collaborations than a single company could possibly provide. This is true for most emerging technologies. Founders of the core competence will need to rely on

external relationships and technologies, identifying applications, market leaders, technology trends/competition etc. and most importantly, stakeholders.

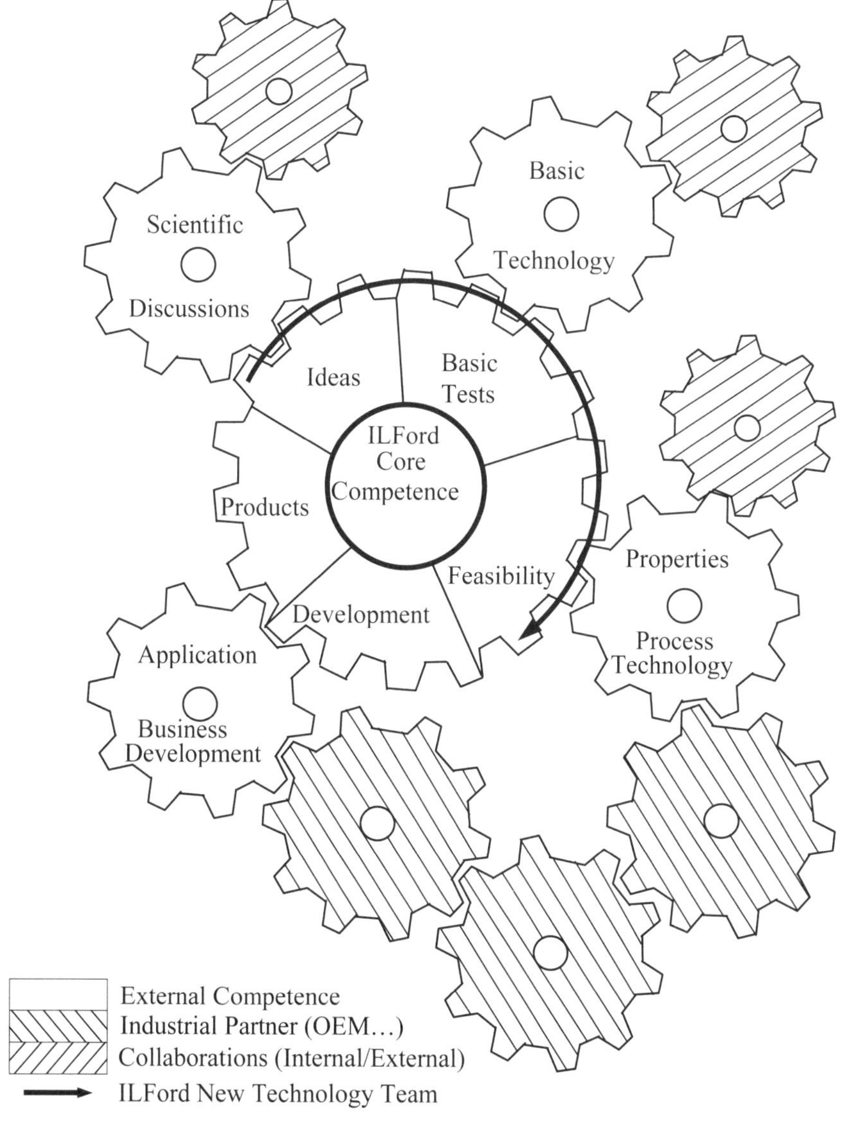

EXHIBIT 2.2 ILFord Innovation Engine

Source: Marc Pauchard PhD, ILFord Imaging Reprinted with permission

The more important feature that Pauchard's diagram demonstrates is the fact that the "Innovation Engine" can only turn clockwise, all the wheels move in their correct directions. The engine must move in a concerted manner, in harmony. We live in an environment comprised of complex surroundings where it is difficult to tell signal from noise. In complex surroundings there is no way to predetermine or schedule innovation. It is in this complex realm where disruptive innovation is born, where continuous learning is a prerequisite. One needs as much stimuli as they can possibly get in order to heighten the chance that they will stumble upon something which will form into patterns. Knowledge is achieved when exposed to as much background noise, or stimuli, as possible, to open up as many opportunities to respond to as possible. Only by obtaining the knowledge to differentiate signal from noise are we positioned to recognize opportunity. By definition, all radical innovations are unexpected results. Defining an initial idea "A" might not lead to the planned product "A", but might lead to discovery "B" and a completely new product "B", as is with the discovery of the Post-It® note.

Post-It® - Unplanned Innovation, long adoption rate

The following excerpt is from *A Century of Innovation, The 3M Story*, published by 3M, 2002.

> In 1968 Spence Silver, a chemist at 3M, was trying to develop a strong glue [Product A]. But his new adhesive was super-weak instead of super-strong [Product B]. It stuck to objects but could be easily peeled off. No one knew what to do with it, but Silver knew he invented a highly unusual new adhesive and didn't discard his new glue. For the next five years, Silver gave seminars and buttonholed individual 3Mers, hyping the potential of this new adhesive and showing samples of it in spray-can form and as a bulletin board. In 1973 with the support of Geoff Nicholson, Silver's new adhesive began to obtain a few small wins for use in products such as bulletin boards in the form of tiles and tapes. Silver's new adhesive continued to struggle for adoption. Adoption came while Art Fry was using pieces of paper to mark the pages in his hymnal. While sitting in the choir loft at his church, Fry turned to a hymn and his scrap paper bookmark fell to the floor. "My mind began to wander during the sermon," Fry confessed. "I thought about Spence's adhesive. If I could coat it on paper, that would be just the ticket for a better bookmark." [Eureka/Causality/Innovation]. Fry went to work the next day, ordered a sample of the adhesive and began coating it on paper. He only coated the edge of the paper so the part protruding from his hymnal wouldn't be sticky. "When I used these 'bookmarks' to write messages to my boss, I came across the heart of the idea. It wasn't a bookmark at all, but a note," said Fry. "Spence's adhesive was most useful for making paper adhere to paper and a whole lot of other surfaces. Yet, it wasn't so sticky that it would damage those surfaces when it was pulled off. This was the insight [Product C]. It was a whole new

concept in pressure-sensitive adhesives. It was like moving from the outer ring of the target to the bull's eye." As Product C moved from the Eureka to development, Fry encountered serious technical problems very early. First, there was the problem of getting the adhesive to stay in place on the note instead of transferring to other surfaces. And, although 3M was known for its coating expertise, the company didn't have coating equipment that could precisely apply adhesive on an imprecise substrate such as paper. It was difficult to maintain a consistent range of adhesion. It would be years before the Post-It® notes adhesive was perfected, prototypes created and the manufacturing process developed. In 1977 3M attempted the first marketing of the product. But customer response was lukewarm at best. After a successful marketing attempt in Boise Idaho, in 1980 3M decided to launch the product nationally. In 1981 Art Fry's Post-It®® Notes were named the company's Outstanding New Product and was awarded the internal coveted Golden Step Award.

The trail of the Post-It® Note's innovation can be traced back to 1968 where the innovation started at the material level. Launch and adoption did not occur until 1980, 12 years in the development and innovation cycle. The final product, Post-It® was not defined in 1968, it was only by combining knowledge obtained and using it to provide a solution to a problem [innovation/intuition] did the final product emerge. Likewise with the discovery of Poly-OLED, the final product was achieved by discovery. Poly-OLED was not a development plan. It was a realized technology only by observing a neighboring experiment. Dr. Burroughes recognized the significance of the discovery. Further requiring validation he violated scientific protocol by ruining another researcher's experiment. Although good research and development need a draft of their projects, one always has to be open to unexpected results.

A former university professor, Dave Perkins, coined the phrase, "Fortune favors the prepared". His message was clear, the better educated the subject, the higher the probability he/she will recognize opportunity when it is presented. Beyond education an individual needs the ability of intuition and be open for opportunity to occur.

When an emerging technology surfaces or warrants attention, what does one do with it? How does one manage it? How does an investor or corporation evaluate it? Or nurture and incubate it? Often emerging technologies are viewed as distractions. Sometimes a "charismatic" individual who represents the technology with a lot of hype accompanies them (note: "Charismatic" is not the same as a "Champion"). Corporations and investors need to use tools or processes for getting past the hype and managing the technology based upon value and technical feasibility, providing stakeholders the ability to make informed decisions.

Valuing Core Competence

When a company out-sources Research and Development (R&D), significant value can be lost. A High-Tech company's success is founded on R&D. Whether this R&D is internal or acquired from an outside source; advances in technologies, or improvements in processes are crucial to positioning a high-tech corporation in the market place. When R&D is effectively managed it becomes an effective intellectual property generator. Intellectual property is critical for market position and strategic control to the organization.

On January 5th, 2004 Planar Systems Inc., a flat panel display manufacturer based in Beaverton Oregon, stock closed at US$25.85 per share, a high for the year. Nine days later Planar announced they would maintain R&D investment at about 4% of overall sales. 4% of sales equates to an R&D budget focused solely on sustaining engineering, limiting research projects to be completed with below zero budget resources. By February 2nd, the stock price dropped to US$16.66 per share. With 14.62 Million shares outstanding this equates to a reduction in their market cap of US$134.4M. This trend continued to May 3rd, where it reached US$11.43 per share further reducing their market cap to US$210.8M. Eleven months later on April 1st of 2005, Planar Systems announced the layoff of their CTO and most of the remaining technical staff. At that time Planar Systems' stock price closed at US$8.50 per share resulting in a new reduction in their market cap of US$253.7M, from January 2004.

The purpose of this exercise is not to identify that a company made poor decisions, but to demonstrate that there is far more to managing a technology based company than value chain management or any singular approach to business models or strategies. Technology can be acquired, if an organization is scouting for it, and the reserves are in place to support such an acquisition. There may be numerous additional factors contributing to the decline of share price for Planar Systems. In the exercise of evaluating their technical expertise, did the executives undervalue their technology position? Did they determine that technology is a commodity readily available for acquisition? Did Planar Systems' management perform a Need, Approach, Benefits, and Competitions based analysis on their value proposition? The continued decline of Planar Systems' market cap can be traced back to the published decision to remove R&D from a technology-based company.

Technology Transfers

Beyond the origins of innovation there are locations where this occurs. As previously mentioned, a Eureka can originate from any number of locations. It may originate from institutions such as a corporation, university, or government laboratory, or even someone's garage. Stakeholder(s), whom control the technology, need to be managed as they attempt to commercialize it.

As universities tend to have an extensive number of discoveries of which an investor or corporation may wish to exploit, the process of technology transfers warrants discussion. Technology transfer processes differ from university to university. Understanding a few fundamental aspects may prove to be beneficial when considering whether to undertake the venture of a technology transfer.

Luis Mejia published a paper in *Technological Forecasting & Social Change 1998*, which provides an overview of the fundamental formats adopted by universities for technology transfers.

There are three basic types of university technology transfer models which have been implemented: Legal, administrative, and market driven. The legal model is composed of a professional staff of lawyers. They do everything from filing the patent applications to negotiating the license agreements. These lawyers typically are not people with business backgrounds.

The second model, the administrative model, generally combines the licensing function with an existing contract administrative office. The main function of this type of office is to process contracts for sponsored research. The staff usually has no technical or business background.

The third model is based on a market-driven approach. In this model, a professional staff that has both a technical background and some relevant business experience manages licensing functions. Relevant business experience would include, for example, having been in product marketing at a high-technology company such as Intel or Microsoft, or an MBA degree. Legal backgrounds are often discouraged in this model. Lastly, an utmost important requirement in the market-driven model is the critical placement of signature authority within the office. This model will not work if licenses cannot be consummated quickly.

This description of models provides an aide when attempting to understand the motives and processes adopted by the university. Incentives and motives of each department within the technology transfer process are not always in alignment. This is frequently demonstrated with delays in signature approvals and renegotiation efforts by different departments upon previously negotiated terms. Siegel et al researched the key organizational issues in promoting successful knowledge transfer and concluded that there are numerous impediments to effectiveness. Impediments are distributed across cultural and informational barriers, and among three key stakeholder types (university administrators, academics, and firms/entrepreneurs), inappropriate staffing and compensation practices in the technology transfer office, and inadequate rewards for faculty involvement.

The schematic in Exhibit 2.3 demonstrates how technology is transferred from universities to industry. The schematic identifies key stakeholders involved in the process, the actions they are responsible for, their motivations and the culture of their organization. This process is generic to all educational institutes along with research centers, and must be understood by industry in developing their understanding of university culture.

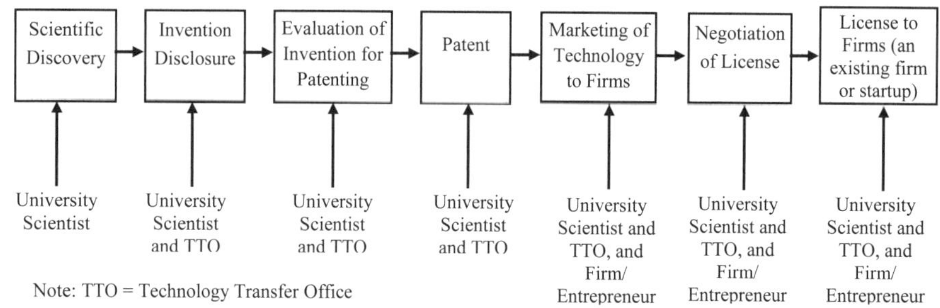

EXHIBIT 2.3 Technology transfer process from a university to a firm or entrepreneur.

Source: D.S. Siegel et al. / Journal of High Technology Management Research 14 (2003) 111–133

Stakeholder	Actions	Primary motive(s)	Secondary motive(s)	Perspective
University Scientist	Discovery of new knowledge	Recognition within the scientific community—publications, grants (especially if untenured)	Financial gain and a desire to secure additional research funding (mainly for graduate students and laboratory equipment)	Scientific
Technology Transfer Office	Works with faculty members and firms/ entrepreneurs to structure deals	Protect and market the university's intellectual property	Facilitate technological diffusion and secure additional research funding	Bureaucratic
Firm/ Entrepreneur	Commercializes new technology	Financial gain	Maintain control of proprietary technologies	Organic/ Entrepreneurial

EXHIBIT 2.4 Key stakeholders in the transfer of technology from universities to the private sector

Source: D.S. Siegel et al. / Journal of High Technology Management Research 14 (2003) 111–133

Understanding the motives behind the three stakeholder types will provide insight into their issues. This knowledge helps identify what agreements should be addressed in terms of content and to which will motivate them. An underlying motive is "Perceived Value", which can be one of the most difficult issues to overcome. Each stakeholder can have very different perceptions of value. It is advised that the individual or organization pursuing a technology transfer be prepared to manage this common and very critical item. As it is also recommended to investigate the university's adopted process to determine if it addresses stakeholder issues and has assigned clear signature authority. Siegel et al provided a table identifying Technology transfer stakeholder motives and is included as Exhibit 2.4

Summary

Adoption of emerging technologies are frequently met with failure or do not achieve expectations. Often the technology can be disruptive or it can enable a completely unrelated or unforeseen opportunity. When the technology becomes mature enough to disclose to selected collaborators, the final application may end up not being what was originally planned. Identifying the required collaborative technologies and the organizations that own them is crucial in the process of moving the technology from the laboratory to the market. Only by understanding the fundamental technology and its role within the technology value chain, can an effective business strategy be architected. Deciding whether to pursue the venture or not depends upon numerous factors all coming together.

Emerging technologies and their applications develop from numerous locations and environments. As with the emergence of the Post-It® and Poly-OLED technologies, they occurred, and therefore were not scheduled. Every emerging technology requires collaborations to emerge from the laboratory to market adoption. They also require time. The Post-It® was in the innovation cycle 12 years prior to adoption and OLED was in it for 15 years. Upon discovery of the emerging technology, the next step is to move it from the laboratory. Moving it from the laboratory can be a simplified or tedious process depending upon whom, or which organization, owns the technology.

Identified by D.S. Siegel et al., universities with established technology transfer processes may have university stakeholder impediments built within their process. This first step of moving the technology from the laboratory requires identifying stakeholders and building relationships with them. As depicted in Pauchard's Innovation Engine, emerging technologies and their applications require collaborations at all levels of development, from fundamental science to market leader adoption.

The effectiveness of moving the technology from the laboratory to achieving market adoption can be greatly improved when one is armed with the tools to manage the supporting channels. Understanding the requirements and functions of the supporting channels will accelerate market adoption. There are additional tools required and available to achieve success. These tools can include key individuals such as Rainmakers and Connectors.

The following chapters describe requirements and functions of the critical aspects for managing an emerging technology. Exampled by referencing case studies, these chapters will demonstrate the types of issues around strategies and technologies, which can occur and require management.

Chapter 3

Technology Development

"...technology development refers to a special class of development projects where the deliverable is a new knowledge, new technology, a technical capability, or a technological platform" - Robert G. Cooper

Technology Value Chain

Emerging and established technologies converge into processes and products designed to provide solutions to market opportunities. There exists a value chain network for each technology as demonstrated in Exhibit 3.1. In this diagram, numerous technologies contribute to the overall value chain enabling function or service, resulting in market revenue opportunities at several points.

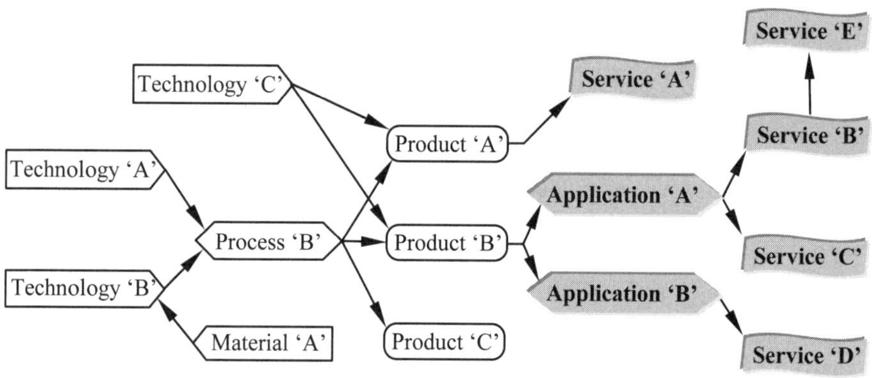

EXHIBIT 3.1 Technology Value Chain Network

Material 'A' can enable technology 'B', which is added to process 'B', resulting in Product 'B'. Products can subsequently combine with other products and processes enabling applications, which can provide service. Services can also provide additional levels of value. An example of additional service levels could be a display product within a mobile phone combined with a GPS (Global Positioning System) microchip. The display provides a human interface function to enhance the user's experience, enabling technologies such as streaming media (Service 'A'). Leveraging the existence of the display and an established customer base, additional advertising revenue could be generated (Service 'B'). The addition of the GPS microchip can provide location services for local merchants and emergency services.

The technology value network is different for every technology. Often, emerging technologies are in search of the "Killer Application". Focus is typically within closely complimenting technologies and processes. Business planning which looks beyond the immediate application to include end service applications can provide additional levels of market value.

As an emerging *material* technology, Poly-OLED required additional technologies and processes for adoption. Exhibit 3.2 identifies, from a high level, different technologies required for adoption within a display application.

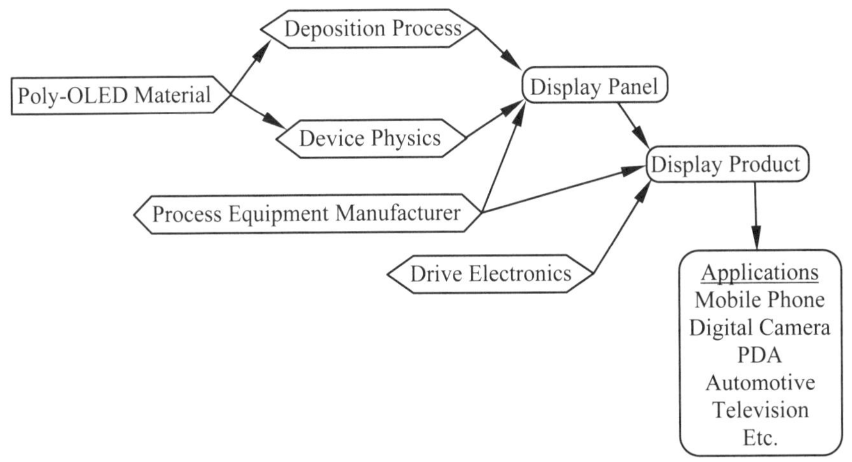

EXHIBIT 3.2 Poly-OLED Display Technology Value Chain Network

Source: Cambridge Display Technologies, Ltd., Cambridge UK

When performing technology development for an emerging technology, the value chain network provides information critical to the decision of, *which technology should be developed internally and which technology can be developed through collaboration or outsourced.* Effective management and execution of this critical decision will enhance adoption and provide market control of the technology. Exhibit 3.3 surrounds the Poly-OLED material with technologies required to create a display product and identifies CDT's decisions on which technology should be developed internally or externally.

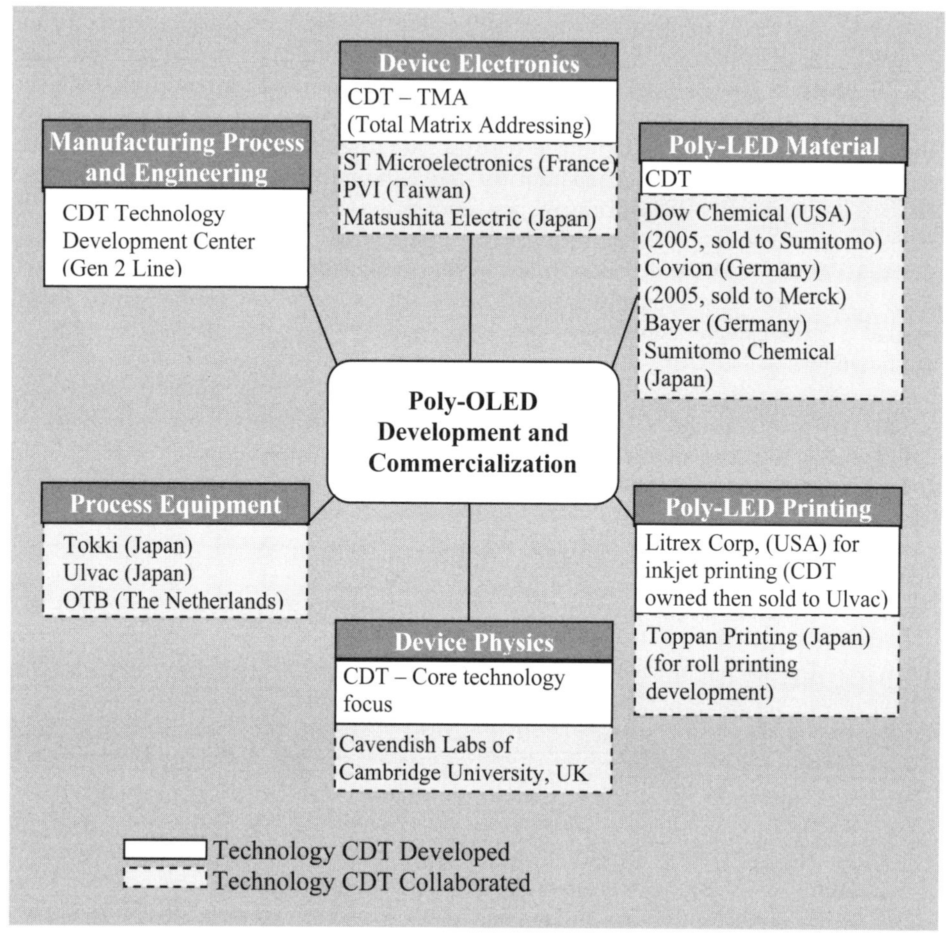

EXHIBIT 3.3 Poly-OLED Technology Ecosystem

Source: Cambridge Display Technologies, Ltd., Cambridge UK (Reprinted with permission)

Excluding development of manufacturing process equipment, CDT managed to maintain a position resulting in both market control of the technology and industry support. CDT held onto critical technologies such as device physics and material development. By collaborating with chemical manufactures, CDT gained accessed to their highly skilled development abilities, whereby accelerating development. One needs to identify an emerging technology's value chain matrix and decide which technology is to be internally developed, collaborated, or outsourced.

The Critical Gap

Innovation initially occurs by observation or through the formulation of concepts. In either case, the next phase is to demonstrate proof of concept. Here in lies the critical gap. Resources are required to demonstrate proof of concept, as the innovator attempts to convince stakeholders to provide these resources. The innovator often struggles to obtain resources to demonstrate proof of concept. In many organizations these opportunities are lost simply because there is no process in place to manage such events. Some organizations have realized this gap and have implemented solutions designed to assist the concept in moving to demonstration (feasibility). Examples of such organizations are the Agency for Science, Technology and Research (A*STAR) Singapore, and The University of Manchester, UK.

Recognizing the importance of commercialization of science and innovation and to boost the economic competitiveness of Singapore, the government of Singapore formed A*STAR. A*STAR is chartered with the task of plotting the future of Singapore's science and technology. It is comprised of a constellation of various organizations such as the Biomedical Research Council (BMRC), the Science and Engineering Research Council (SERC), Exploit Technologies Pte Ltd (ETPL), the A*STAR Graduate Academy (A*GA) and the Corporate Planning and Administration Division (CPAD). Both BMRC and SERC promote, support and oversee the public sector's R&D research activities in Singapore. A*GA supports A*STAR's key thrust of human capital development through the promotion of science scholarships and other manpower development programs and initiatives. ETPL manages the Intellectual Property created by the research institutes and facilitates the transfer of technology from the research institutes to industries. The corporate group comprises functions such as Finance, Human Resource, Corporate Communications, Legal incubation and spin-off management.

The University of Manchester UK, identified the gap between discovery and proof of concept, and established The University of Manchester Intellectual Property Limited (UMIP). UMIP works with researchers throughout the university evaluating and providing seed funding to demonstrate proof of concept. UMIP provides additional resources, in a phased approach, to the opportunities as they achieve milestones. This program can continue through the establishment of the corporation, assistance with startup funding, startup management and incubation facilities. This process nurtures the innovation process through commercialization providing required assistance along the development path beyond filling the gap.

Some organizations simply provide an incubation solution. Both A*STAR and The University of Manchester have gone beyond filling the gap between discovery and proof of concept. They have provided the required business and legal aspects most organizations overlook.

Application Specifications

Business development will identify and analyze several sectors for which the technology can be applied. These will be presented in a chart similar to Exhibit 3.4 identifying market sector, cost benefit and volume.

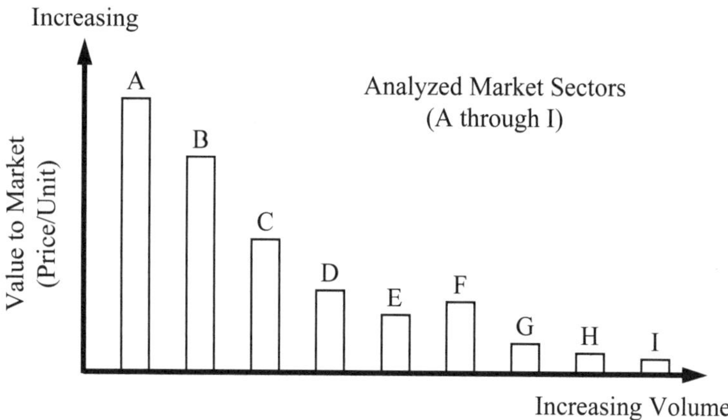

EXHIBIT 3.4 Technology Market Sector Value & Volume

There exists a general trend of increasing volume resulting in decreased incremental value to market. Large volumes demonstrate a market accepted solution. These accepted solutions are not experiencing much pain and therefore are not highly motivated to change or alter the currently adopted solution. Whereas lower volume applications tend to be in a continuous search for better solutions and are more willing to pay for improvements. They are also more apt to accept not so strict performance specifications for early stage technologies.

Each sector will have defined specifications required for adoption. The emerging technology will align better with some sectors rather than others. As the technology development progresses, it is better to focus on lower volume higher value added sectors, rather than chase the lower value high volume (commodity) markets. This approach would generate revenue and provide the time to refine the process and technology developments required to meet the demands of the higher volume markets. This approach was also exemplified by Geoffrey A. Moore's, Crossing the Chasm.

This analysis assists the organization in strategic planning when the required specifications for each market are compared to the specifications of the emerging technology and its ability to achieve these specifications. It also serves as a staged product planning tool for future development and growth, focusing only on those sectors with higher revenue and adoption probability.

Stages of Technology Development

Exhibit 3.5 provides a high-level view from an operational and business perspective of the typical stages of development for emerging technologies. An important metric for research and development departments is the generation of intellectual property. Also included are the different types of intellectual property commonly generated during varying stages of technology development.

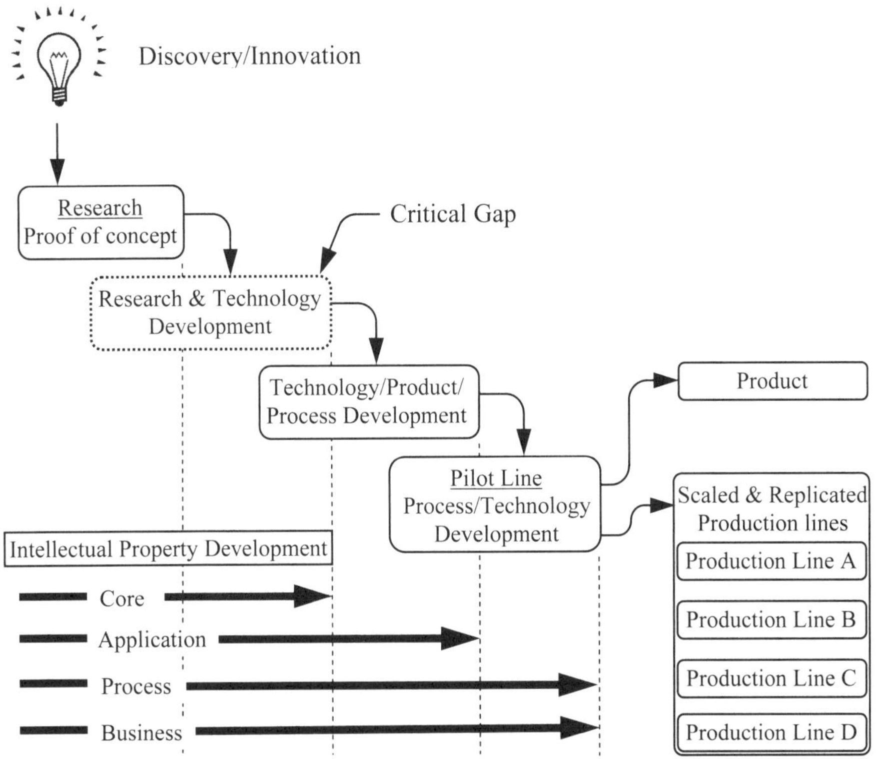

EXHIBIT 3.5 Stages of technology & intellectual property development

In terms of a *material* innovation or discovery such as a new chemical formulation, the first step is discovery. This is when laboratory experiments are performed to demonstrate proof of concept, typically in a beaker. The technology will subsequently undergo stages of technology development as it progresses beyond the beaker to production. It will then require technical characterization. This is when the technology undergoes a comprehensive list of experiments and modifications to develop knowledge about the specification limits of the technology. With the technology characterized it can then progress into a product development cycle where it is designed into a product. Additional specification knowledge is learned as the technology undergoes minor modifications to function within the application. With final design approval the technology can migrate to a pilot line. The pilot line is designed to focus on processing or manufacturing parameters of the technology. The output of the pilot line can be either a product targeting a market, or intellectual property and knowledge transfer.

In an intellectual property and knowledge transfer model, the organization certifies the pilot line under their control. It is then duplicated through licensing agreements to other companies. The other companies refer to the pilot line as the master line where improvements are developed over time and subsequently transferred to the duplicated/licensed production line. Market size and strategic control are determining factors when deciding if licensing the pilot line is viable strategy. A small volume market may not need more than one or two production lines to meet demand. While a very large volume market may require several production lines to meet demand. The organization owning the technology ensures that the number of licensed production lines is limited to avoid commoditization of the technology.

Proof of Concept

When innovation occurs, it initially exists only in thought. As with intellectual property applications, the concept must be reduced to practice. This is achieved by either documenting how this can be performed within the patent and/or performing development upon the concept. Often, this first level is simply a sanity check. A fundament experiment can be performed to provide validation before moving on to the next stages of development.

Early levels of development can also be achieved through the use of external contract engineering organizations. Contracting some levels of development to external companies can yield the required product. This approach delays the capital requirement of hiring the technical staff and equipping them with the required development tools.

Industry leaders for each market sector will often demonstrate skepticism about the claims or promise of an emerging technology. These leaders are approached many times from different directions and technologies, each making promises and claims of better or improved performance. Due to the nature of the market, this predisposed disposition is justified. It is up to the emerging technology to differentiate itself and demonstrate its abilities. One of the easiest methods to demonstrate the claim is to create a functional prototype. This prototype occasionally requires development outside the scope of the business model. For example, a material manufacturer fabricating Quantum Dots claims that the material can be used to create an emissive display. This will be met with a list of known technology issues in manufacturing from the display industry. The material manufacturer is not in the display business nor has plans to build displays but is now required to demonstrate proof, hence create a display. This can be achieved through either collaboration or the creation of a micro fabrication line. All that will be required is a device that addresses the most common list of concerns. A fully functional video display is not required. Both solutions of collaboration or micro fabrication are viable and have pros and cons, Exhibit 3.6.

Collaborations allow for reduction in short term expenditures with limited effect upon continuing burn rate, yet expose the organization to a higher risk of losing strategic control. Investing in a micro fabrication line increases short term and continuing operational burn rate, while providing abilities to generate additional IP and maintain control. The pivotal factor is the confidence of the organization's ability to locate a collaborator and the associated risk.

Business Impact	Collaboration	Micro Fabrication (organic development)
Initial cost	Low	High (capital)
Intellectual Property	High Exposure Possible loss of control	Maintains control Generates additional IP
Personnel	Legal & Executive time No increased investment	Additional resources, increased investment
Facilities	Little or no effect	Increase required Increased investment
Ability to produce additional samples	Often slow	On demand
Probability of achieving	Varied	High

EXHIBIT 3.6 High level assessment of organic development verses collaboration

A prototype or functioning device has the benefits of exciting investors, future customers or potential collaborators. Each organization will have different strengths, competencies, and business model concerns which will determine whether to invest organically or collaborate.

Collaborations

Know and manage your strengths, outsource or collaborate everything else. Collaborators can come with many benefits. They can serve as an extension of R&D resources, enabling access to technologies and expertise that would otherwise be cost prohibitive to startup organizations. They can also provide access to market channels and early adoption to products. Aligning with the appropriate collaborator can improve the emerging technology's industry reputation on viable feasibility and improve investor confidence.

The appropriate collaborator is defined as a strategic collaborator, not necessarily the first company who knocks on the door asking to collaborate. When analyzing the technology value chain for a specific application, market leaders with complimentary technology will be identified. These market leaders are preferred collaborators. They will have channels and access to other channels through their existing collaborations.

While building collaborations keep in mind that every organization is working toward their strategic road map. These projects are budgeted annually and large corporations are typically not in a position to drop everything and change direction. They will not have engineers and scientists sitting around waiting to work on your technology. However, when large a organization decides to collaborate they will internally spend several multiples of capital ($) on the development than the startup organization will or could.

Occasionally, startups approach a tier one organization with a desire to collaborate, and expect the tier one company will pay the startup a US$100K to US$1M collaboration fee. This message to the tier one organization is that the startup does not understand the value brought or the expenses that will be incurred by the tier one organization. It also sends a message that the investors of the startup do not support the technology and will not fund such developments. Concerns over long term stability arise and reduces interest in the emerging technology. Collaborations must be managed carefully. Relationship management skills across multiple cultures prove to be a very valuable asset.

Product Development Processes

Many successful companies adopt a product development process similar to the Project Flow Chart identified in Exhibit 3.7. Each organization will refine or tailor the process to function within their organization. This is accomplished by creating a product development process-networking model depicting all the departments involved in the product development process, all the tasks they perform and the information flow required among departments. Phase or Gate reviews are scheduled at critical junctures to determine if the project remains in alignment with the corporation's goals and objectives. Is it on track with the original specification and scope? Has the market changed requiring an alteration of product specifications?

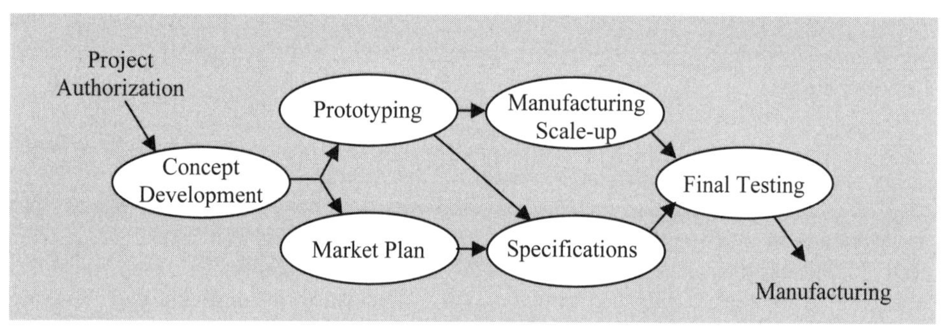

EXHIBIT 3.7 The Project Flow Chart

Source: Getting the most out of your product development process.
Harvard Business Review March–April 1996

Product development processes of this nature have proven to be effective for *established* organizations. However, these organizations can still struggle while attempting to manage the gap between discovery and product development. *Product development* combines a collective of defined processes and technologies to create a product for market. *Discovery* requires additional research to determine all of the technology's characteristics. A technology should not be placed into a product development process until all of its characteristics are defined. If this occurs, the unknown characteristics result in development delays and increased costs (delays to market, etc.).

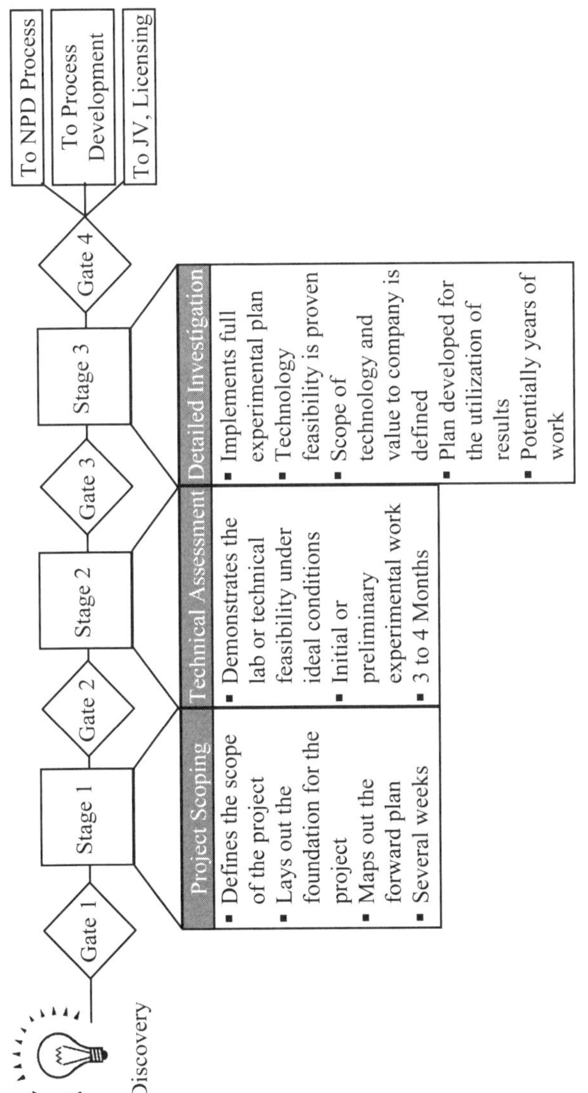

EXHIBIT 3.8 Technology Development Stage-Gate® Process

Source: Product Development Institute, Inc., 2007

For many technology startup companies, and Technology Development (TD) discoveries within an established organization, Robert Cooper recommends readdressing the Stage-Gate® process and reducing it to 4 gates and 3 stages, Exhibit 3.8. The process identifies periods of development (Stage) and periods for review (Gate). This refinement can include calculation of Return On Investment (ROI), firm identifications of risks, projected sales and a host of other quantitative factors. For early stage technology development, this information is a guess at best, commonly resulting in a "kill" decision due to unknown variables.

Management of the product development process is a different task. Project management is more than simply driving a gaunt chart of scheduled tasks. It involves managing items such as cost, quality, risk and communications. In 1987 the Project Management Institute published a white paper in an attempt to document and standardize generally accepted project management information and practices. The first edition of the PMBOK® (A Guided to the Project Management Body of Knowledge) was published in 1996 and is currently an internationally recognized standard (IEEE Std 1490-2003). It documents the fundamentals of project management as they apply to a wide range of projects, including construction, software, engineering, automotive, etc. The Guide recognizes 44 processes that fall into five basic process groups and nine knowledge areas that are typical of almost all projects.

Adopting Product Development and Project Management processes allows the organization to capture and manage all processes required to effectively drive a technology to production. Leveraging these existing knowledge bases help identify all processes that need to be accomplished within its discipline.

Research & Technology Development

Transitioning a technology from a one time demonstration in a lab to a micro development and characterization process is often attempted with a limited budget similar to many universities. For example, a researcher determines they can ink jet print conductive materials to fabricate a transistor. Often the researcher will purchase a commercially available desktop ink jet printer and modify it to perform experiments. On the surface this sounds like a practical approach and occasionally works out fairly successful to demonstrate a one time sample. However, what frequently occurs is that the researcher learns there are many different types of ink jet technologies and there is a large amount of chemistry involved; from material compatibility through an ink jet head, chemistries for the material to be ink jet printed, chemistries on the substrate, and secondary processes required to anneal or cure the ink jet printed material. Performing this development process on a modified desktop printer will yield significantly different results than performing them on an industrial ink jet printer in a clean room. The controlled environment of the industrial equipment in a clean room will allow better repeatability. Better repeatability equates to shorter development time and the ability to refine processes to improve yields. However, there is a significant difference in cost when one moves from a US$30 ink jet printer to a US$1M industrial ink jet printer placed in a class 1000 clean room.

Many emerging technologies evolve from universities and as such, come with the low budget university research mentality of performing development with the lowest possible equipment and environment cost. This is a paradigm that when overcome can transcend a university startup from a "Great group of researchers" to a strategic organization. With the proper equipment product development can focus on achieving specifications required by the market, not on modifying a broken ink jet printer with limited functions. They can implement Design of Experiments (DOE) procedures developed by authors such as Genichi Taguchi and Douglas Montgomery. Development and process intellectual property will also surface, providing additional value to the organization. The technology will be closer to a real production process rather than having to redevelop the knowledge when production equipment eventually arrives.

This is achieved by creating a properly equipped micro laboratory which purpose and function is to refine and characterize the technology, its associated process, and generate intellectual property. Creating a manageable transition from laboratory to fabrication (lab-to-fab) is achieved by implementing proven processes with proper equipment in their appropriate environments.

Product Development Planning

The creation of a marketing plan and a product development plan, mark the beginning stages of a product development project. These documents are reviewed and authorized (Exhibit 3.7). Authorization of these documents serves as a kick-off to product development and deployment of resources. These documents provide clear communications to stakeholders in terms of what tasks will be performed, over what time frame, to achieve a specific result, at a realistically estimated cost. These are very helpful when managing expectations and ongoing changes to scope or markets.

Summary

Technology development starts with identification of an application. The technology is then tested and adjusted to achieve the specifications required for that application. Initial applications often conceptualized in the beginning may not be the preferred application for rapid adoption. Performing analysis on applications from different sectors will help determine which is in better alignment with the technology and which provides the higher value. Then the technology value chain for that sector can be analyzed to identify potential *strategic* collaborators.

As the technology undergoes migration from the lab to refinement for a specific application, there exists a critical gap. It is here where the technology needs funding support to perform specification characterization. This can be performed internally or outsourced to a contract development organization. The Product Development Institute recommends a 4 gate and 3 stage process designed to capture the critical aspects while providing the ability to manage tasks and expectations.

Getting the technology across the critical gap can require knowledge beyond the startup organization. Collaborations can provide access to development resources beyond the capability of the startup organization. When viewed as an extension of research capabilities, collaborations can help the emerging technology achieve specifications required for adoption. Stepping out of the "Develop at lowest possible cost" model to a "Properly funded and equipped development" model will increase probability of success and reduce development time.

A technology should not be placed into a product development process until all of its characteristics are defined. If this occurs, the unknown characteristics result in development delays and increased costs (delays to market, etc.). With the technology defined, utilization of proven product development practices captures the required tasks to be performed from a comprehensive organizational perspective; it improves communications and manages expectations.

Chapter 4

Strategic Marketing

"Dozens of studies over the years have found the most significant predictor of success of a new product is the degree to which it is different from existing options. It pays to be patient and persistent in developing products with superior benefits and points of differentiation" - David A. Aaker

What is the Business Strategy?

"What is the business strategy?" this is the first question often asked when an emerging technology is initially disclosed. This serves two functions; first, it determines how much business planning has been performed, and secondly, when viewed as a distraction, the question is designed to discredit the technology before resources are allocated or diluted. Research around the technology needs to be performed to analyze applications and *ALL* requirements to achieve marketability. Only then can one ask, "What is the business strategy?"

Strategy

Strategy is an overall plan architected to obtain a preferred market position. A tactic is a scheme designed for a specific action within the strategy. A solid and sound strategy is derived from a comprehensive analysis of the technology, its value proposition to the customer, and the organization's effectiveness to deliver and execute. This analysis includes understanding market dynamics, cultural impacts, competition, and government regulations. Developing strategies without this knowledge is similar to playing chess with your self. You are guaranteed to win!

As emerging technologies begin to find their application and route to market channels, the market dynamics can appear to be comprised of a chaotic cloud, Exhibit 4.1. Adoption rate can be improved by systematically analyzing this chaotic cloud using a structured approach to identify a preferred target market segment and differential advantage. Strategic marketing identifies the initial direction to start the customer discovery process.

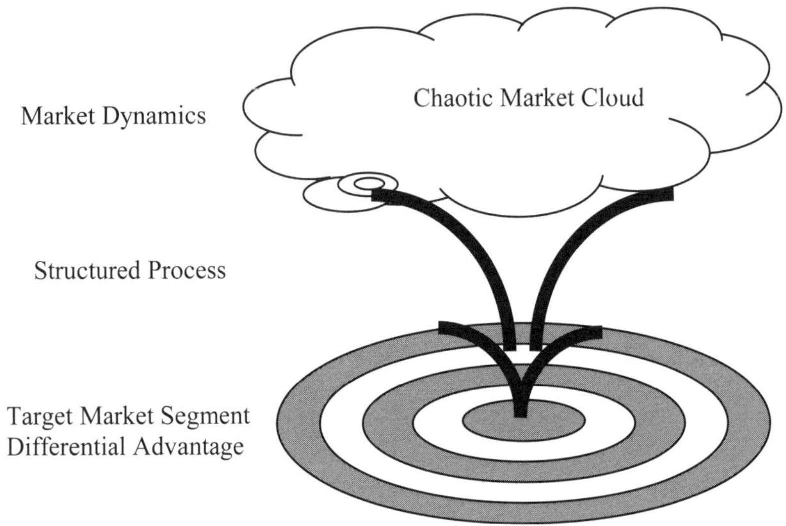

EXHIBIT 4.1 Path to Target Market Segment and Differential Advantage

Strategy is to be aligned with the corporate objectives. Corporate objectives should be more than "Obtain X% Market share". Market share does not necessarily return *value* to share holders. The objective should be based upon shareholder return on investment. For example, ROIC > CC (Return On Investment Capital > Cost of Capital) or "Maintaining a rate of profit above industry and competitive levels" are objectives which align shareholder interests with corporate objectives. Building a strategy which supports the objectives starts with defining the objective(s), analyzing the industry and organization, then finalizing with decisions and execution, Exhibit 4.2.

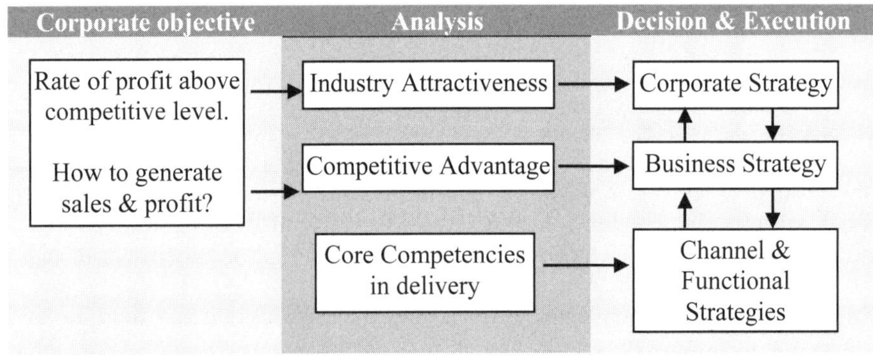

EXHIBIT 4.2 Path from corporate objective to channel and functional strategy

There exists numerous market analysis and business analysis processes, all have pros and cons. The important aspect is to implement a structured framework designed to analyze market data, organizational capabilities, and technical value chains resulting in a clearly defined business strategy. The following presented processes are designed to provide such a structured framework to define the target market segment and differential advantages.

SWOT

Mitigating weaknesses and threats, exploiting opportunities and strengths: before a firm can define its target market segment and differential advantage it must know itself and how it relates to the market. What does the firm do better than others (strengths)? Where does the firm need improvements (weaknesses)? By performing a SWOT (Strengths and Weaknesses, Opportunities and Threats) analysis these factors will be revealed.

A SWOT analysis is performed along the common line of the value chain, Exhibit 4.3. Strengths and weakness are to be analyzed upon the firm (internal), while opportunities and threats are analyzed upon the market (external).

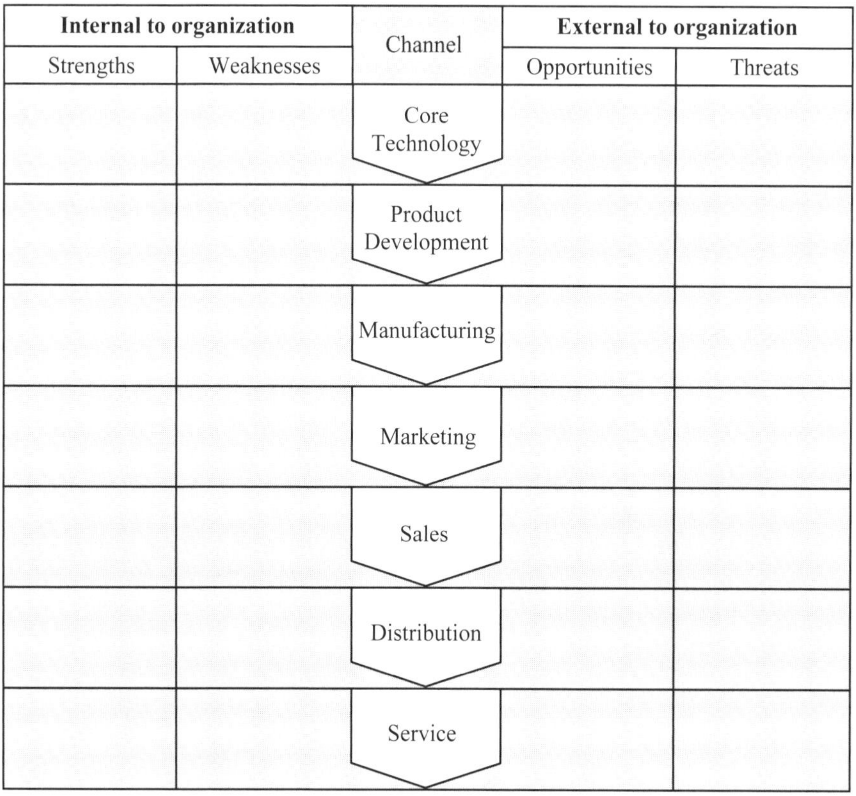

Internal to organization		Channel	**External to organization**	
Strengths	Weaknesses		Opportunities	Threats
		Core Technology		
		Product Development		
		Manufacturing		
		Marketing		
		Sales		
		Distribution		
		Service		

EXHIBIT 4.3 Organizational SWOT analysis

As the management team performs the analysis, identification of weaknesses within the organization will emerge. These may require mitigation through additional investment or through collaboration and outsourcing. The firm might be strong with their core technology but weak in product development. Product development requires developing products that adhere to government regulations and engineering processes. A decision must be made to either outsource product development or hire experienced staff.

Identifying strengths can provide opportunities for exploitation of the differential advantages. Does the firm possess proprietary manufacturing methodology which reduces time to market, provides a substantial reduction in costs, or is it technically advanced such as first generation semiconductor manufacturing? These attributes provide a differential advantage that can be exploited to penetrate, protect, or increase market share.

Emerging technologies can be driven from a startup or organic organization. An organically driven technology can leverage the organization's channels, brand, resources, etc. Where as startup has not brand, or infrastructure to leverage. They must develop the infrastructure to build a business. By performing a SWOT analysis the management team will identify which areas need mitigation and which need exploitation.

Macro and Micro approach

A high level to bottom level approach starts with an analysis of macro factors which effect market dynamics. These macro factors are political, economical and regulatory (government) in nature. Government regulations can prohibit adoption by restricting the use of heavy metals in consumer products (RoHS Regulations, European Directive 2002/95/EEC), or enhance market adoption such as the Food and Drug Administration's (FDA) mandate to use RFID (Radio Frequency Identification Device) on all drugs.

Cities and municipalities in the United States are beginning to prohibit full motion video on billboards due to their impact on increased traffic accidents and congestion. These restrictions include limiting digital billboards to displaying pictures for a specific number of seconds, removing video. Additionally, technologies enabling video tend to be significantly heavy and require structural improvements. Some areas of the USA billboard market are prohibiting structural improvements and/or new billboard structures. This reduces the adoption of video and places pressure to use printed signage, or develop lighter weight technologies. Governmental regulations can provide both market leverage and market restrictions.

In this global economy, one must analyze each region's governmental regulations. This can expose regions that are better positioned for adoption and some which have regulations prohibiting adoption.

Michael Porter 5 forces

Michael Porter identified 5 market forces which impact a firm's potential performance, Exhibit 4.4. An environmental analysis should include dynamics within these five forces. This model provides a framework for analyzing the competitive strength and position of a corporation or business organization. At the center lies the *competitive rivalry* comprised of entry and exit barriers, geographical factors, incumbent's resistance, new entrant strategy, and channels to market. This center is influenced by external forces from: Powerful *Suppliers* who can exert influence on an industry when; they are concentrated, there are significant costs to switch suppliers, they possess a sole source position, or when they possess a credible forward integration threat. Suppliers are weak when buyers are concentrated, there are multiple sources for buyers or when the market is comprised of large cooperative purchasers. *Buyers* are strong when they are concentrated; purchase a significant amount of output or posses a backward integration threat. Under these conditions the buyer sets the price. Buyers are weak when they are fragmented or when producers supply critical components of the buyer's offering, producers can take over or develop their own channel, or when products are not standardized resulting in high buyer switching costs.

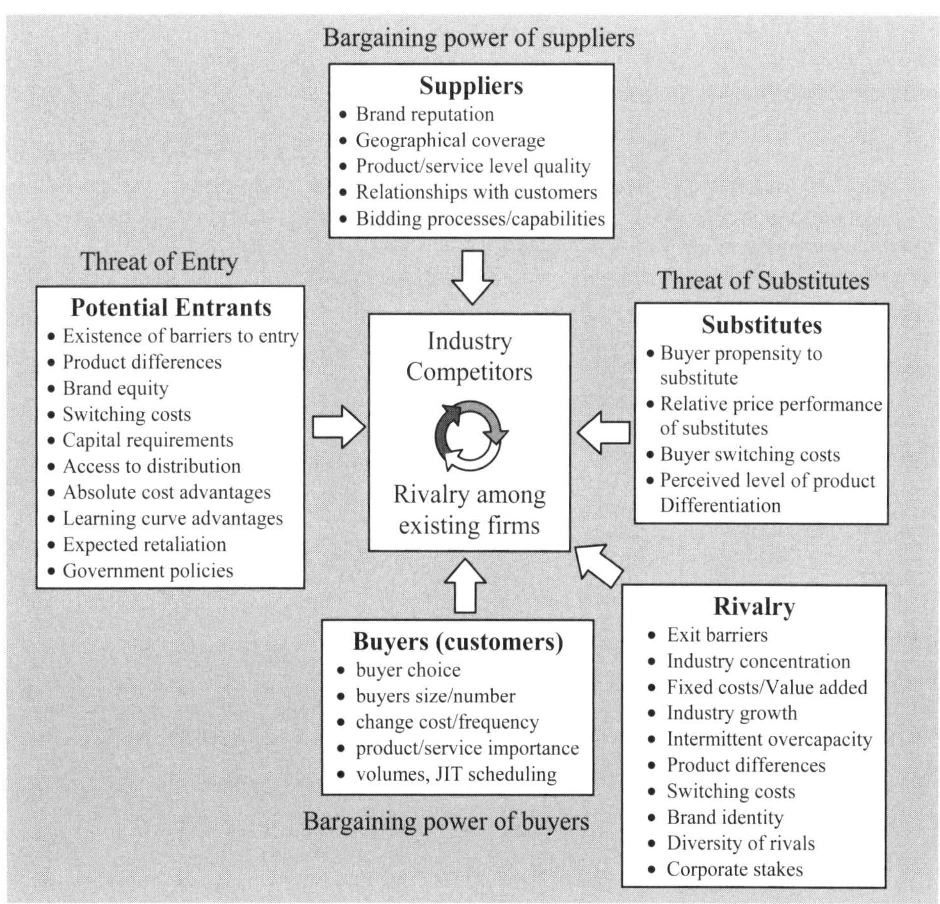

EXHIBIT 4.4 Porter's Five Forces of Competitive Framework

Source: Porter, Michael (1998) Competitive Strategy: Techniques for analyzing Industries and Competitors, The Free Press

Threat of *Substitutes* comes from products outside the industry. A product's price elasticity is affected by substitute products - as more substitutes become available, the demand becomes more elastic due to customers having access to more alternatives. A close substitute product constrains the ability of a firm to raise prices. The threat of *Potential Entrants* largely depends on the barriers to entry. Barriers are high if a market requires highly intensive capitalization or the return on capital is low (commodity market). Government regulations could prohibit or enhance competition. Markets and other entrants can retaliate or be resistant to a new entrant.

Incumbent rivals are not the only threat to firms in an industry; the possibility that new firms may enter an industry also affects competition. In theory, any firm should be able to enter and exit a market, and if free entry and exit exists, then reasonable profits should always be achieved. It is in the "Threat of Potential Entrants" where an emerging technology attempts to enter the market. Success will depend upon comprehensive market knowledge, execution of strategic planning and value proposition.

Market Evolution (BCG strategic environment matrix)

As forces impact market dynamics the market will evolve from a fragmented condition to an eventual commodity market. The Boston Consulting Group (BCG) represented this migration with a graph aligning the number of differentiators to the value of the differentiators, Exhibit 4.5.

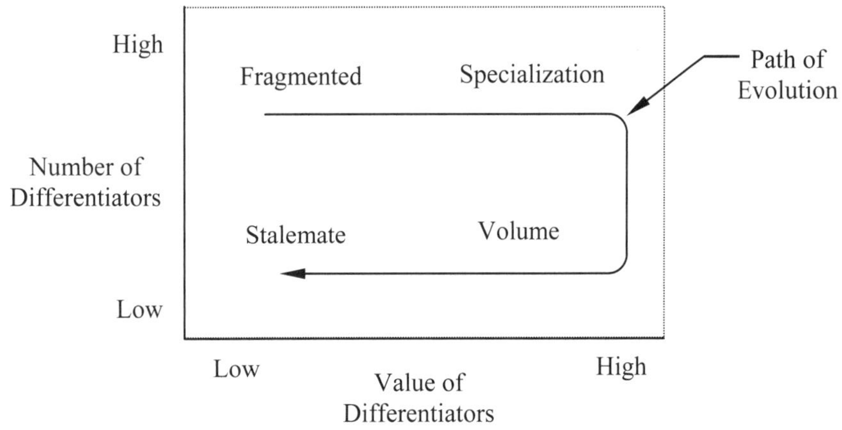

EXHIBIT 4.5 Industry Evolution - Strategic Environments Matrix

Source: Boston Consulting Group

Emerging technologies can enter the market at any stage of evolution. A new market can be developed resulting in the fragmented stage. As an example, assembling

Light Emitting Diodes (LED) to create video signage. Similarly, an emerging technology can enter a commodity market by providing a lower cost manufacturing solution resulting in an increased competitive advantage.

Exhibit 4.6 identifies the market conditions and success factors for each stage of the evolutionary process. Profits decrease as the market evolves toward stalemate. These are critical considerations to be evaluated while understanding the market and determining strategy.

Stage of Evolution	Market Conditions	Success Factors
Fragmented	• Differentiated products • Low brand loyalty • Diffused technology • Small scale economies • Channels are diffused	• Low costs • Novel Forms of differentiation • Quick response to change
Specialization	• Varied Customer needs • First mover advantage • Brand loyalty • Economies of scale	• Environmental Stability • Ability to systemize
Volume	• Reduction of players • Increasing volumes • Clearly defined market sectors	• Size • Production scale
Stalemate	• Highly competitive • Similar strategies • No significant advantage • Commodity market	• Early identification of stalemate • Timely exit strategy

EXHIBIT 4.6 Industry Evolution – Condition and Success Factors

Source: Boston Consulting Group

Gather market data

Emerging technologies can target more than one market for adoption. For example Poly-OLED can be exploited in the display market, photovoltaic markets, laser printing markets, lighting and communications to name a few. While analyzing which market to pursue it must to be divided into sectors. Sectors are to be defined along channels to market, not necessarily along technology or product lines. Channels are aligned with customer requirements for specific market sectors and define specifications and/or requirements for adoption. Sector analysis also identifies tier structure and opportunities for penetration.

There are several firms which collect and analyze market data. Their business models are comprised of selling standardized market reports and performing custom research. It is recommended to leverage these firms to obtain market data. As an example, iSuppli located in California USA divides the signage and professional display market into nine sectors. Exhibit 4.7 identifies these sectors, their value and projected growth trends. For a company developing an emerging large screen display technology such data would prove to be valuable.

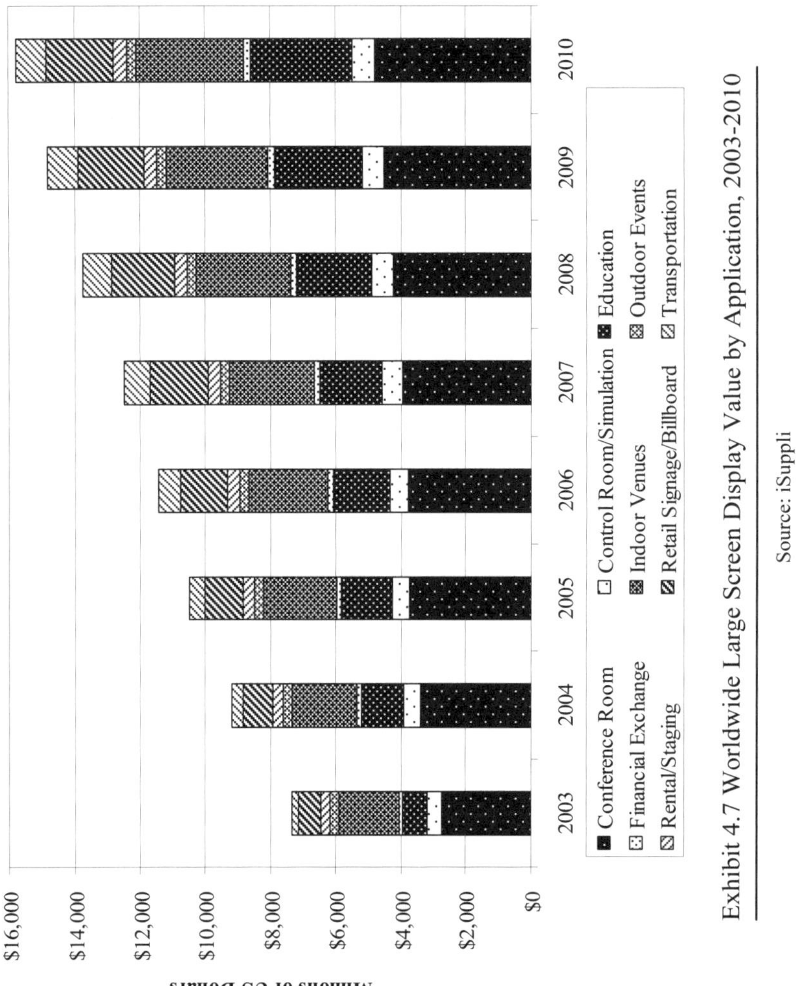

Exhibit 4.7 Worldwide Large Screen Display Value by Application, 2003-2010

Source: iSuppli

Additional technology and market data can be obtain by attending a few industry conferences and performing searches on the Internet. Internet search engines may need a little assistance when it comes to cultural differences. Care must be taken when performing searches. To find a report on "commercialization" in Australia or Europe, it is suggested to change the spelling to "commercialisation".

The data within exhibit 4.7 will be used to demonstrate the preceding examples of analyzing market attractiveness and competitive strength. Raw market data depicting sector definitions, CAGR and revenues by technology can be found in Appendix A, tables A.1 through A.11.

Market Attractiveness and business's competitive strength analysis

With the market and subsequent sectors identified the next step is to analyze the data to determine which market segment is best positioned for successful adoption of the emerging technology.

The Mckinsey approach position's a firm's SBUs in a 3 x 3 matrix according to their market attractiveness and the organization's competitive strength, Exhibit 4.8. The major application of this matrix is to help managers align the firm's strengths with the available market opportunities. The firm should align the emerging technology and organizational strengths with market sectors that are most attractive.

Firm's competitive position \ Industry attractiveness	High	Medium	Low
High	1	2	4
Medium	3	5	7
Low	6	8	9

Variables that might be used to evaluate:

Firm's competitive position		Industry attractiveness	
Size	Distribution	Size	Price levels
Growth	Technology	Growth	Profitability
Relative share	Marketing skills	Competitive intensity	
Customer loyalty	Patents	Technical Sophistication	
Margins	Brand image	Government regulations	

EXHIBIT 4.8 Market attractiveness and business's competitive strength matrix

Source: McKinsey and Company

Shell Chemicals uses a similar model that it refers to as the "directional policy matrix", and assigns the following strategy recommendations to each of the nine cells:

1. *Leader.* This is the optimal position for an SBU, a strong position in a highly attractive market. The strategy is to provide top priority to enhancing or maintaining this position.
2. *Growth leader.* Investment should be made to enable the product to keep pace with market growth.
3. *Try harder.* This position might be vulnerable over time. Consideration should be given to investment to strengthen its position over time.
4. *Cash generation.* These SBUs should be cash suppliers and should not require investment.
5. *Proceed with care.* Caution is required when investing in these businesses, since neither are they market leaders nor are their markets strikingly attractive.
6. *Double or quit.* Businesses here should be decisively partitioned into those to be abandoned and those selected for priority investment.

7 and 8. *Phased withdrawal.* Profit prospects are slight and the strategy should be a controlled switch of resources to other opportunities.

9. *Withdrawal.* These businesses will be loosing money and their assets should be disposed of as quickly as possible.

A firm's preferred position is Cells 1, 2 and 3. These represent a strong alignment between the firm and the market. To determine cell location of market sectors for an emerging technology, the following *Sector attractiveness* and *Competitive advantage* analyses must be performed.

Sector attractiveness

With a background understanding of market attractiveness and the firm's competitive position, the management team gathers to perform the following analysis. It is designed to align the corporate strategy with the market and identify strategic market segments resulting in a faster rate of adoption.

To demonstrate this process of sector analysis for attractiveness, data for the large screen display market will be used, Exhibit 4.7. Market sector definitions, growth rates, volumes and revenues will also be used. The management team develops a list of variables which are important to the firm and applies a weighted significance to them. The list of variables from Exhibit 4.8 for industry attractiveness can be leveraged and expanded upon. Exhibit 4.9 compiles the data into a simplified format with selected variables and weighted significance applied to them. Through discussion and debate the management team then assigns a rating of high, medium, or low to the sectors and tabulates final scores.

Sector Attractiveness Ratings

High 10 Medium 5 Low 1

Weight	Attractiveness Factors	Indoor Venues	Outdoor Events	Retail/Signage/Billboard	Conference Room	Education	Transportation	Control Room/Simulation	Financial Exchange	Rental Staging
		Sectors								
20%	Growth Rate	5	1	10	5	10	10	5	5	1
20%	Sector Volume	10	1	5	10	10	5	5	1	1
15%	Channel (Stable/Fragmented)	5	5	10	1	5	5	1	5	1
15%	Acceptance of Technology	10	10	5	5	10	5	5	1	5
20%	Standardization	5	5	10	10	10	5	5	10	1
10%	Service	5	1	5	10	10	5	10	10	1
100%	**Final Scores**	6.8	3.8	7.8	6.9	9.3	6	4.9	5.2	1.6

EXHIBIT 4.9 Market attractiveness analysis matrix

Competitive advantage

The same process is repeated for the competitive advantage matrix, Exhibit 4.10. Again, the list of a firm's competitive position variables in exhibit 4.8 can be leveraged and expanded upon. These are variables the management team believes are significant to a successful organization. The data for Exhibit 4.10 is filled in from a position of an emerging technology within a startup organization. Typically startups can not leverage brand and they do not possess a sales/Service/Marketing infrastructure, resulting in low ratings for those variables. The opposite can be true for organically grown emerging technologies.

Organizational Strength Ratings

High 10 Medium 5 Low 1

Weight	Organizational Relative Strength	Indoor Venues	Outdoor Events	Retail/Signage/Billboard	Conference Room	Education	Transportation	Control Room/Simulation	Financial Exchange	Rental Staging
		Sectors								
20%	Sales Potential	10	5	10	10	10	1	5	1	1
20%	Cost Competitiveness	10	10	10	10	10	10	10	5	5
15%	Sales/Service/Marketing Capacity	1	1	1	1	1	1	1	1	1
15%	Brand Image	1	1	1	1	1	1	1	1	1
20%	Market Share	1	1	1	1	1	1	1	1	1
10%	Technology	10	5	10	10	10	5	10	5	10
100%	**Final Scores**	5.5	4	5.5	5.5	5.5	3.2	4.5	2.2	2.7

EXHIBIT 4.10 Organizational strength analysis matrix

Target Market Segment

Due to continuously changing market conditions and numerous other factors, managers will have varying opinions on the high, medium and low ratings. The objective is to gather a unified agreement from the management team. Then place the scores in the market attractiveness and business's competitive strength matrix defining which market segment should be targeted and why. Exhibit 4.11 compiles the data to reveal the firm should pursue indoor venues, retail/signage/billboard, conference room, and education sectors. These are the target market segments for the firm as derived from the process. Utilizing such a process develops strategy from an analytical approach and removes the "Gut feel" approach. The matrix is indicative of startup organizations indicating a weak infrastructure. Further investments in infrastructure (sales, service & marketing) can be focused around targeted sectors pushing some of them into a highly competitive position. This can be achieved through strategic hiring and collaborations.

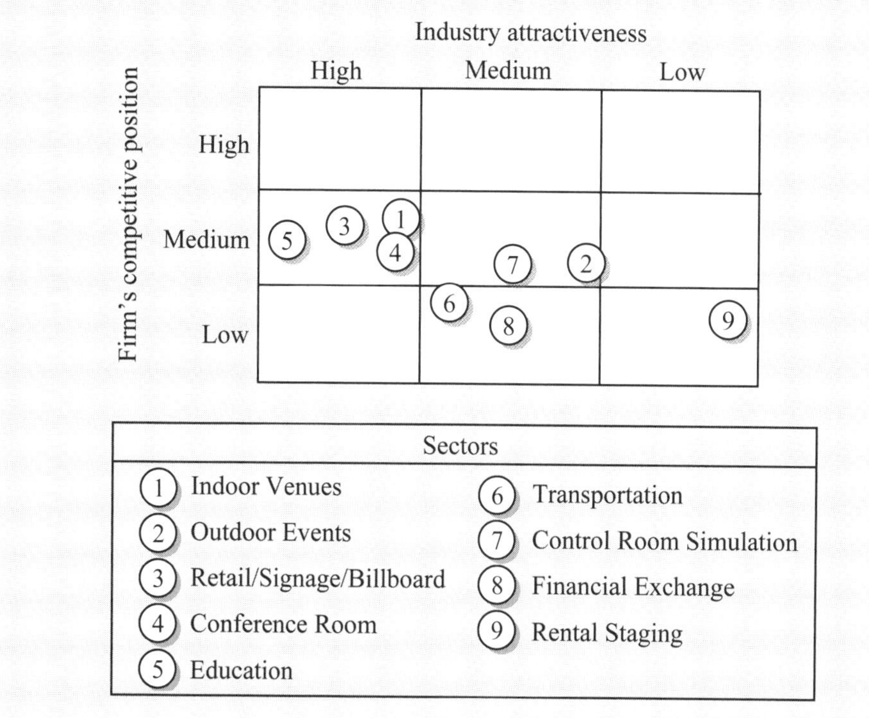

EXHIBIT 4.11 Market attractiveness and business's competitive strength matrix

Value Proposition & Differential Advantage

Each sector has product or service specifications defined by the market. A market adopts an emerging technology only if it fulfills the need. Under this premise product specifications are to be defined by the market. Comparing these product specifications to the results of an application and value chain analysis for the emerging technology reveals the value propositions and differential advantage.

Application Analysis

An application analysis studies the market for not only growth sectors, but trends in technology, identification of competing technologies, and the trends of industry leaders. Understanding the fundamental function of the technology will aid in the identification of other emerging and adopted technologies, which provide the same or similar function. These technologies may, or may not, be equivalent but they must be acknowledged and understood. What technologies are currently being used by the industry? To what sector(s) are these technologies migrating? Observing industry leaders and their technology investments will aid in providing insight to answer these questions. Where is adoption most likely to occur and for which application? What collaborations are required to enhance the adoption rate? Which companies are best positioned to be a strategic collaborator?

As described by Pauchard and demonstrated with the 3M Post-It® story; emerging technologies can be adopted in unforeseen applications. Usually these patentable applications are in unrelated market sectors. One must make a proactive effort to identify applications through discussions with individuals who possess varying backgrounds. Another example is Quantum Dot technology. This technology emits a fixed color of light (red, green, blue, etc) when an ultra violet light (UV) is shined upon it. Technically obvious applications are for displays and photovoltaic applications. Perhaps an easier and possibly faster adoption application could be cosmetic for products such as mobile phones. Most phones already contain a UV light source which could activate a clear plastic case to emit color patterns when opened, or patterns when exposed to black light rooms as found in social settings. One must consider markets of vanity and youth, for these applications can have rapid adoption with the highest return.

Value Chain Analysis

Analysis of the value chain will demonstrate where the emerging technology fits within existing applications and most important, how it is differentiated. Comparing advantages and disadvantages between similar technologies will clarify the emerging technology's value proposition and its value to the organization (internal and external). The emerging technology should be mapped and compared to existing technologies to communicate a clear message of its value proposition. Tabulate differentiators, technical, operational, financial, and use specifications, against competing technologies. Similar to the Application Analysis section a SWOT analysis is to be performed from the Value chain perspective. The SWOT analysis for both sections can be performed in a single iteration, on the condition that it includes perspectives of both Value Chain and Applications.

Three aspects of the value chain are to be analyzed to develop a comprehensive list of value added differentiators, Exhibit 4.12. This analysis compares the emerging technology against the top 2 market leading technologies.

Value Chain Factor	Differentiators
Market Value	What is the economic value to the customer resulting in enhanced performance? Lower cost? Improved experience? Increased efficiencies?
Technical Value	Lower power? Higher performance? Improved physical characteristics?
Operational Value	Cost less to manufacture? Reduction of environmental impact? Require less energy? Smaller production facility?

EXHIBIT 4.12 Value chain differentiator factors

As another example of value chain analysis Exhibit 4.13 demonstrates the technical value differentiators between a market adopted active-matrix liquid crystal display (AM-LCD) and an active-matrix OLED display (AM-OLED).

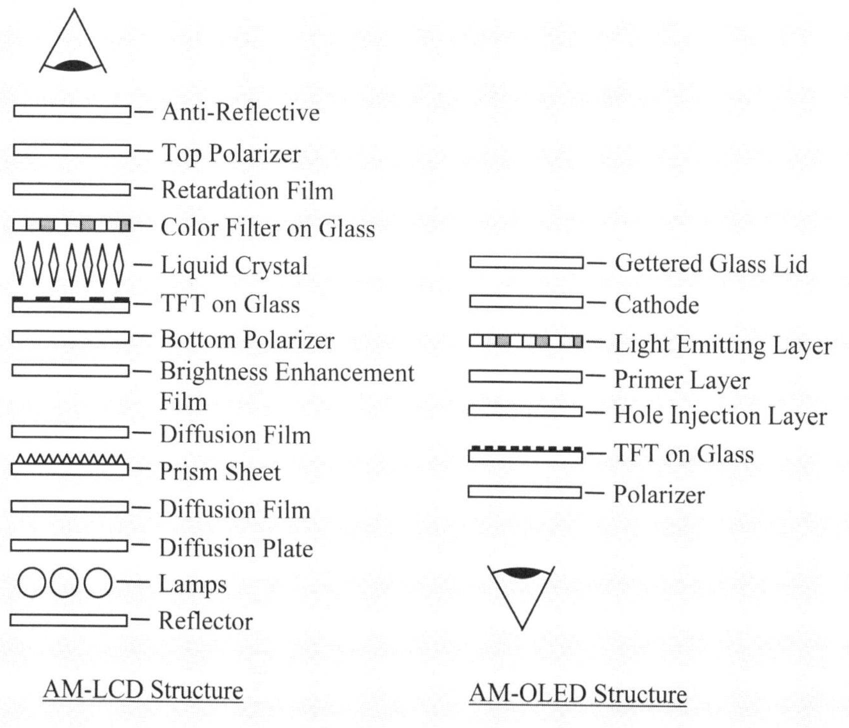

EXHIBIT 4.13 Comparison of AM-LCD and AM-OLED device structures

In this example the AM-OLED display has a simple structure compared to AM-LCD, with a 50% reduction in components. One can observe that the emerging AM-OLED technology can be easily adopted within established production lines at a lower cost, and can utilize key components of the existing manufacturing infrastructure.

For an emerging technology to be adopted it also needs to provide a cost justification. Can the technology be manufactured and, if so, what are the operational costs? This is to be a detailed analysis, not an estimate. If the emerging technology is attempting to displace an adopted technology, what impact does it have on the adopted process? This analysis is to include the actual process to manufacture the technology and the improvements upon the utilization of the technology in its application. This will provide numbers for cost reduction in materials, capital equipment, time, facility, CO_2 reduction, environment impact, personnel, etc.

An often over looked value position is application value or more recently termed economic benefit. Occasionally, the management team can be too focused on ROI overlooking an enhanced value to the customer. To demonstrate this Exhibit 4.14 analyzes the ROI for conversion from printed graphics to digital signage for POP (Point Of Purchase) Posters in three use categories.

ROI Model for Ink-On-Paper to Digital Signage Conversion			
Cost (POP Posters)	**High User**	**Average User**	**Low User**
Printing (number of Signs)	$52,840 (60)	$10,368 (30)	$2,160 (10)
Installation	$2,880	$576	$120
Lost Signage	$10,368	$2,074	$432
Total			
Per year	$65,088	$13,018	$2,712
Over 5 years	$325,440	$65,090	$13,560
ROI (achieved in 1 year if $Digital Sign <)	**$1,000**	**$425**	**$250**
ROI (achieved in 5 years if $Digital Sign <)	**$5,250**	**$2,000**	**$1,350**

1) Average Display size 18 sq-ft.
2) Analog Cost per sq-ft (US$2.00)
3) Cost to install each poster (US$2.00)
4) Number of display turns/year (24, 9.6, 6)

EXHIBIT 4.14 Digital Conversion ROI Model for POP Posters, Worldwide

Source: IT Strategies, Boston MA, 2006

The analysis above indicates that if the digitally controlled sign cost is $1,000, a high volume user would achieve their ROI in 1 year, for 60 sign locations. Should this be a target product selling price for an emerging digital signage technology? In the September 8th issue of BusinessWeek Magazine the two following quotes where published.

"In one case in Pittsburgh, 11 Lamar-owned boards were generating only $11,000 a month in revenue. After being switched to digital technology, they're pulling in more than $50,000 a month each."

"JCDecaux uses scrolling billboards to display up to six ads each, targeted at different times of the day. The result is that scrolling boards generate as much as €15,000 ($19,100) a day, on average, vs. €5,000 for a static board."

In one case, Lamar increased generated revenue by 50X with digital conversion, while JCDecaux increased revenue by 3X after converting to scrolling boards. This data represents the *economic benefit* of the proposition of converting a static advertising sign to a digital sign. An emerging technology has the possibility to alter a customer's business model. This possibility should be considered and leveraged accordingly when deciding upon the selling price (value proposition) for the business model.

Competition	Market Drivers	Market Inhibitors	Opportunities & Threats
New & Existing Entrant Forces	• Value added • Market size	• Intellectual property • Channel relationships	**Opportunities** • Collaboration • Acquisition **Threats** • Relationships • Technology
Supplier Forces	• Concentration • Product differentiation • Cost conditions • Existing players • New entrants	• Technology o Cross platform o Cross process	**Opportunities** • Suppliers entering market • Conglomerations **Threats** • Potential emerging technologies
Buyer Forces	• Increased price sensitivity • Regional coverage • Sectors • OEMs • MNCs vs SMEs	• Buyer formed cooperatives	**Opportunities** • Increasing supplier market forces • Fragmented channels (acquisitions) **Threats** • Cost pressures
Substitution Forces	• Market size	• Substitute adoption • Brand loyalty	**Opportunities** • Collaboration • Licensing **Threats** • Lower cost solutions accelerate time to commodity market

EXHIBIT 4.15 Other market drivers analysis

Other market drivers

As Porter describes in his 5 forces model there are 4 sections which impact the market. *Potential Entrants* is where an emerging technology affects market dynamics. There are often several emerging technologies attempting to position themselves in a state of adoption. Each technology will be positioning their value as superior on several differentiators. One must not overlook other emerging technologies or readily dismiss them.

Perform a high level analysis on the market. Determine market drivers and inhibitors for each of the four market forces. Identify opportunities and threats for each.

Exhibit 4.15 provides an example analysis of other market drivers. Additional items may apply to the targeted market.

Risk Management

Throughout the development of the emerging technology, associated risks will change. Additional risks will be added, some will diminish and some will be significantly altered. Utilization of FEMA (Failure Effects Modes Analysis) and risk profiling techniques for impacts to the technology can be used to monitor and adjust the strategy as required.

Risk profiling includes identifying sources of risk arising from the market, government regulations, technology, and organizational capabilities. These risks are to be listed in a Risk Matrix with a significance value and proposed mitigation assigned to each. Exhibit 4.16 provides a sample Risk Matrix of items from each risk dimension. The list will change or grow to accommodate the emerging technology.

Risk Dimension	Risk	Rating	Mitigation
Government	Regulatory Cultural Business Design Liability	High Medium Low None	
Market	Competing Technologies Collaboration Development Business Design Cultural Blocking Strategies	- or -	
Organizational	Distribution Channel Dependence Upon External Resources Speed of Change Available Capital Management Structure		
Technology	Maturity IP Portfolio Alignment with Core Competence Manufacturability Projected Cost Supply Chain (Maturity/stability) Competing Emerging Technologies	1 2 3 4 5	

EXHIBIT 4.16 Phase 2 Risk Identification Table

Source: Slywotzky, Adrian and Morrison, David (2001) *The Profit Zone.* Three Rivers Pres

Every risk requires acknowledgement and comment for the purpose of generating awareness, monitoring, and determining mitigation. Throughout the process, newly identified risks will emerge and are to be added to the matrix.

Product solution(s)

Make it easy for the customer to adopt the technology. This is achieved by understanding how the technology fits within the customer's value chain and packaging the technology in such a way as to conform to adopted processes and standards. Prototyping the technology will demonstrate feasibility and provide opportunities for additional IP. Customers will demonstrate greater interest when they observe technical maturity.

Business Model Profile

Analysis of the technology and its value chains will generate early indicators of an initial business design. This is when high-level business strategies are developed. It is imperative to follow a standard process for developing a business strategy, ensuring all aspects are properly addressed. The business development model adopted here is an expansion on Slywotzky et. al's standard business design dimensions as depicted in Exhibit 4.17.

Section	Key Dimensions
Customer Selection	• Immediate Customers (OEM) • Application Developers • End Users
Value Capture	• Profit Sales • Value Sharing • Solutions/Service
Differentiation/ Strategic Control	• Price • Technology • Intellectual Property • Specialization • Global Network • Brand • Customer Relationships
Scope	• Products • Solutions • International • Collaborations
Culture	• Educational Plan • Language/Terminology Barriers • Government Regulations

EXHIBIT 4.17 Phase 2 Business Design Dimensions

Source: Slywotzky, Adrian and Morrison, David (2001) *The Profit Zone.* Three Rivers Pres

Each section of the business model is reviewed by the management team. The business design dimension of Culture has been added to capture possible solutions to issues related to culture in this growing global economy.

Resource Plan w/financials

Managing an emerging technology to market adoption will require a plan with resources. A business strategy coupled with technology development will require investment for execution. The previously analyzed Market attractiveness and competitive strength analysis identified areas requiring investment. Chart these areas, correlate investment to strategy and timelines for milestones, Exhibit 4.18.

Milestones are to be aligned with the strategy. Each section may need to hire staff and/or consultants to achieve milestones.

Channel	Strategy	Timeline (months)	Required Investment	Responsible Person
Research & Development	(Milestone(s) achieved by required resources)			
Operations				
Marketing				
Sales				
Distribution				
Service				

EXHIBIT 4.18 Investment strategy and milestone delivery matrix

Summary

The market, originally appearing as a chaotic cloud has been broken down using a structured approach. In conjunction with the executive team, the target market segment, value to the market, differentiation, and requirements for the organization to achieve adoption have been defined. When moving the technology from the beaker to the market requires defining these aspects. What has been presented is a collection of business processes organized into a structured format designed to help an organization determine where they are, where they want to go, and what's required to get there. The collection of business processes utilized are only a small sample of the numerous available to the reader.

Reviewing the previous business strategies and business planning processes, one can determine that the process of performing strategic business planning for emerging technologies can be very complex and enduring. Frequently termed "Managing Innovation", the in-depth knowledge and perseverance required to successfully manage an emerging technology to the point of industry adoption requires a comprehensive understanding of multiple aspects and can be achieved with a good team which possesses the ability to make the tough decisions and understands the tenacity of the work required.

As demonstrated throughout this book, creating the Eureka in a laboratory is only the beginning. In order to achieve market acceptance a much larger team with very valuable skills is required. Managing an emerging technology to achieving market acceptance is a path filled with failures and successes, both major and minor in nature. The comprehensive knowledge required is not held by one person and is best served by formulating a highly skilled team. Success will be determined not only by adoption but how long it takes to achieve adoption. Time for adoption can be reduced by following the structured process outlined in this chapter. It ensures management will investigate and make informed decisions, which are commonly overlooked and can result in problems of delay with marketplace adoption.

Strategic Marketing should focus on reaching a deep understanding of customers and their problems. Analyze the early adopting customers to discover what differentiates them, resulting in the creation of a repeatable road map of how and why they buy. Appropriate milestones for measuring progress, as defined by Steven Blank are:

1. How well do we understand what problems customers have?
2. How much will they pay to solve those problems?
3. Do our product features solve these problems?
4. Do we understand our customer's business?
5. Do we understand the hierarchy of customer needs?
6. Have we found visionary customers, ones who will adopt the technology early?
7. Is our product a "Must have" for these customers?
8. Do we understand the sales road map well enough to consistently sell the product?
9. Do we understand what we need to be profitable?
10. Are the sales and business plans realistic, scalable, and achievable?
11. What do we do if our model turns out to be wrong?

Technologies and how cultures use them are continuously changing. Pay attention to the cultural aspects or use-models of the business strategy and be prepared to change the entire organization to accommodate these changes. Look for the "Quick win", the easiest application that can adopt the technology. Be prepared to perform the development required that will enable the customer to utilize it. This may require additional funding from stakeholders. Communicate to stakeholders what customers require and what is required to achieve their adoption. Stakeholder management is a never-ending process and must be continuously performed.

Chapter 5

Business Development

"Unfortunately inventors and the authors who write about them often underappreciate the importance of innovators and the larger team required for achieving marketplace success" – Innovation, Curtis R. Carlson and William W Wilmot.

"It pays to be patient and persistent in developing products with superior benefits and points of differentiation"- David A Aaker. A lack of patience and persistent usually comes from stakeholders such as executives, investors, inventors and suppliers whom have not learned the highly valuable lesson with respect to this statement. Managing an organization for charisma, while driving it for short-term gains often fails to create a sustainable corporation. Too often executives managing a technology-based company will attempt to redefine the business model from a technology solutions provider to a pure service provider. This transition from technology to service is often accompanied with the removal of research and development investments, under the belief that IP can be easily acquired. This transition can attempt to place the organization into the most profitable position in the value chain while exposing it to value chain pressures of cost reduction. As value chains are constantly evaluated and squeezed for reduction and optimization, an organization seeking alignment into the most profitable link could readily find themselves with a rapidly declining market share; as emerging technologies and service models are constantly evolving.

Business Strategy

Occasionally a company's business strategy may not be conducive to market dynamics. For example attempting to control and generate *ALL* IP for a technology while attempting to grant non-exclusive licenses to everyone. Licensees can push for exclusive rights, while licensors only want to grant non-exclusive. One can find themselves negotiating this opposing interest while attempting to maintain a positive relationship, resulting in a driven forward opportunity. Balances are often found by providing time-limited exclusive rights.

Some companies try to execute a business model of charging exorbitant fees for simply reviewing a technology; performing the due diligence required to manufacture and sell products based upon the technology. This initial fee for reviewing has planned follow up fees for manufacturing, selling and distribution. These fees are in addition to ongoing licensing fees. Somehow these organizations believe that a company will not perform due diligence prior to spending US$100K or more. Their justification is founded on the premise, "If it is important to them, the fee is insignificant and demonstrates the customer's intent. We only want to deal with serious companies". This type of business model appears to have the potential of generating lots of revenue on fees alone, prior to licensing. All the company needs is 10 interested companies paying US$100K each to investigate a technology and they would generate US$1 Million. Looks good on paper but the market is rarely sold on such concepts. A company performs comprehensive technical due diligence *prior* to making any fee investment. This is similar to frequently stated business strategy justifications for China, "There are 1.3 Billion people. All we need is a small percentage to be successful!" Of course those who perform business in China will conclude that a more successful approach is to understand that China is made up of 10 provinces with 13 million people each, and 10 distinct cultures.

Beyond the value chain; knowledge of the market, its changing dynamics, cultural and corporate relationships helps provide insight to develop a sound business strategy which has a higher chance of market adoption.

There exist two approaches to business development; 1) Find another customer (sales), and 2) align the business to market expectations. Finding another customer for an existing product is better described as identifying emerging market opportunities. Successfully aligning a business to market expectations requires a different approach and knowledge base. Provided by examples, and followed up with aspects of communication development, the following outlines value added abilities and knowledge commonly found in successful business development professionals.

Cultural Impact & Educational Planning

An often over looked and under valued aspect of a business strategy is the impact of *Culture*. Considering the cultural impacts of one's strategy is critical, and when overlooked can prove to be detrimental. Performing strategic business planning across the numerous cultures of this global economy is not a meager task. After World War II, the United States was clearly the manufacturing champions of the globe. As time passed, standards of living increased, cost of labor increased, and worker's rights grew. The United States manufacturers started to look west for lower cost solutions. This migration west started with an expansion of manufacturing in Japan. Japan assimilated manufacturing practices from the United States and perfected them, developing processes such as Continuous Process Improvements (CPI), Statistical Process Controls (SPC), and Design of Experiments (DOE) to name a few. Penetrating the Japanese market proved to be a daunting task as a result of protectionism and government controls. In turn, Japan started to experience the similar increase in cost of manufacturing, increased workers rights, and an increased standard of living. As a result they too looked west toward Taiwan P.R.C. and Korea. This eastward trend has continued for decades, and will continue. As the economy of each country improves, so does the standard of living, thus opening markets for expansion. Each country has a different culture with varying differentiations of government regulations. Ignoring and disrespecting these cultural and governmental differentiations, or assuming they are the same as one's own country could prove to be fatal. There are a surprising number of business strategists who use the phrase, "China has 1.3 billion people, and all we need is a fraction of a percent to be successful. We can't go wrong!" As a result there is an ever-increasing list of business strategies for China, which have resulted in complete failure. Successful companies rapidly observed, learned and adopted their strategy to the different culture.

Cultural – Across Borders

Amway has an example of how culture can impact a business strategy and how the organization adapted. Amway has a 100% satisfaction guarantee policy. If the consumer is not 100% satisfied with the product, Amway will replace it free of charge. Gan Chee Eng, Vice President (Greater China Region) of Amway China Company Limited, described the situation as Amway brought their guarantee to China;

> "I tried to explain to corporate that their guarantee will not work in China, but they insisted. People would have a wagon in the parking lot with a small barrel in it, come into the shop and purchase a 1 liter container of L.O.C.™, walk out to their wagon, dump the container into the small barrel, walk back inside, and say, "I'm not satisfied, you replace" Honoring the guarantee almost put us out of business on the first day. We closed for two weeks and re-opened with a new guarantee, which limited customer satisfaction to providing one replacement, which meant we effectively sold two for the price of one."
> - Gan Chee Eng

Culture aspects of business designs are not limited to cross-cultural differences such as taking a proven business design from the United States and expecting it to be just as effective in a different country. Cultural aspects also include how people, corporations, and governments interact within an industry. The government of China required Amway to have a storefront. Multi-level distribution business designs where an unacceptable form of business in China and Amway had to redefine their business model. After opening with a storefront, Amway discovered that their 100% satisfaction guarantee required a modification. By October of 2004 with continuous refinement of their business model and strategy, Amway China Co. Ltd achieved sales past the $2 Billion mark. Macro economical forces (cultural and governmental) impact on market opportunities and business strategies must be observed, for ignoring them can prove to be detrimental.

Cultural – Macro Effects

Government and industry mandates can impact the success of emerging technologies. Radiofrequency Identification tags (RFID) are commonly used in retail applications, mostly for inventory and to prevent shoplifting. They are also widely utilized in the security industry for employee badges. RFID is comprised of a small computer chip connected to an antenna. When the antenna receives a signal from a transmitter, the chip returns a signal containing the identification of the device, which corresponds to an employee or product number. When responding to an emerging threat of drug counterfeiting the Food and Drug Administration (FDA) released a report in July of 2004, which stated that “Radiofrequency Identification (RFID) tagging of products by manufacturers, wholesalers, and retailers appears to be the most promising approach to reliable product tracking and tracing…and that it is feasible for use by 2007”. This report has been interpreted by many as the “FDA RFID mandate” requiring all drugs to be RFID compliant by 2007. In conjunction, the large retailer Wal-Mart announced in June of 2004 that it would require its top 100 suppliers to put RFID tags on shipping crates and pallets by January 1st 2005. In July of 2004, Wal-Mart announced that it would expand its RFID efforts to its next 200 largest suppliers by January 1st of 2006. These are a couple examples of the impact government and industry can have on an emerging technology. With both sectors creating mandates for the use of a technology; investment, expansion, and adoption certainly follow.

Cultural – Across Industries

Cambrios is a startup company founded in 2004. Cambrios currently uses wet chemical production methods to fabricate nanostructures. Originally the company was founded on the basis of using genetically modified proteins with affinity for electronic materials to fabric such nanostructures In October of 2004, Cambrios hired Dr. Hash Pakbaz from the high-tech industry as Vice President of Business Development. Having the pleasure of sitting in on one of Dr. Pakbaz’s early presentations, I watched as he started out explaining how Cambrios’ technology worked. His presentation included graphical slides designed to provide clarity to the technology and how it can be applied to semiconductor fabrication. Imagine yourself as an individual containing in-depth knowledge of semiconductor processes, sitting through a presentation containing a vocabulary of words such as affinity, peptides, virus and bacteriophage, which are not used to describe semiconductor processes. In short, Cambrios, before moving to wet chemical manufacturing techniques used proteins to assemble inorganic Nano particles into controlled shapes. Proteins have specific locations, which bind to other items such as proteins, materials, etc. By isolating a protein, which binds one end only with glass and the other end only to silver, Cambrios can use this protein as a building block to coat silver onto glass. This technique can be expanded to three-dimensional structures, as is performed in nature. I asked Dr. Pakbaz, if he believed individuals in the semiconductor industry understood his presentation? Dr. Pakbaz replied, "After the presentation they just sit there quiet with their eyes glazed over. Then they reply with, "Show me a transistor and its electron mobility". I do not have the impression they understood the fundamental science. They just want to see a transistor." Is this an example of culture difference? Cambrios was combining the science from two separate industries with completely different vocabularies. Positioning, building collaborations, and obtaining early adoption for this emerging technology will not be an easy task. Cambrios faced a situation where stakeholders are not aware they are stakeholders and Cambrios has to convey the value differentiation of their technology. In order for Cambrios to get their message across, an educational plan, which addresses vocabulary differences, had to be developed. Cambrios needed to create a bridge between the biotech culture and the semiconductor culture. Additionally, Cambrios had to demonstrate the fundamental claim of their technology by creating a uniform surface coating, testing its adhesion and electrical properties, creating a transistor, and testing its electrical characteristics. Achieving these two milestones would provide answers and numbers required by the semiconductor industry. It would also improve the probability of adoption. Assuming an industry leader of one technology sector will completely comprehend the technology of another technology sector only leads to disappointment, confusion, and frustration.

Cultural – Social Economic

Reported by the World Health Organization (WHO) in 1998 and reproduced in a publication by the *Health In Your Hands* organization in 2002, The three leading Infectious Killers of children was acute respiratory infections including pneumonia and influenza, AIDS, and diarrhea diseases, in that order. With diarrhea accounting for 2.2 million deaths, of which approximately 1.6 million were children under the age of 5[7]. The solution to effective intervention for reducing infectious diseases is hand washing. It is estimated that if hands were washed systematically with soap over 1 million lives could be saved each year. Taking an example from a case study for the Bottom Of the Pyramid (BOP), more recently referred to as the Bulk Of the Population (BOP). In C. K. Prahalad's book, "The Fortune at The Bottom of the Pyramid", he makes reference to and provides an example of "Stimulating Demand Through Education" via a case study on Hindustan Lever Limited (HLL) the largest soap and detergent manufacturer in India. By performing extensive market research, reformulating the Lifebuoy soap and implementing an educational program, HLL observed an increase of 30% in Lifebuoy sales. The market analysis revealed that consumers were not using soap because they did not believe their hands were dirty, or that using soap provided any added value. HLL developed an educational program based upon "*visual clean is NOT safe clean*", which included a "*germ-glow*" demonstration targeted at school children ages 5 to 13, and to their parents. C. K. Prahalad included an important quote worth mentioning:

> "In the communication we are just speaking about the category of soaps. We are advocating soap usage, we are not advocating Lifebuoy. But the branding, the elements that we put around it, are all branded with LifeBuoy." - Harpreet Singh Tibb at HLL

This quote demonstrates HLL's commitment to society and to their corporate values, which are an integral part of their business strategies.

Cultural – Historical

Another lesson on how business strategies can benefit from cultural knowledge can be learned from the origins of the sailing industry. A Sextant means "Six" tant. It divides the world into base six. It actually measures the degrees from a stellar object relative to the horizon. For example the North Star, also called "Polaris" meaning the star over the Pole, sits directly over the North Pole. If you were on the North Pole a sextant would read 90 degrees when looking at the North Star. If you are on the equator, the North Star is right on the horizon so you read "0". All the measurements in between provide your latitude position (the horizontal circle around the earth in degrees). It wasn't that many years ago when they called protractors "compasses" in school. It is because they literally ARE compasses without the magnetic needles in them. The compass, trigonometry, time and all units of distance are based on navigation and the units are divisions of the circumference around the world. The sailing legend is that base 6 numbers were used to keep the math relatively simple, but hard enough so that Captains would have control over their crews. Once at sea the captains were the only people who could get the crew home, the sextant and navigation math was like the first software security password. Here is how the units work:

> A compass has 360 degrees (4 x 90 for each direction N, S, E & W). Latitude and longitude (horizontal and vertical divisions of the earth) are divided into these 360 degrees.
>
> Each degree is divided further into 60 minutes - so 360 degrees x 60minutes
> = 21,600 minutes around the world = 21,600 nautical miles (sometimes called "knots"). A minute of degree and a nautical mile are the same. Even today they don't put legends for distance on nautical charts, just the latitude and longitude lines.
>
> Time is the timer for the worlds rotation - 360 degrees in 24 hours = 15 degrees an hour (easy math, but not too easy). So at 15 degrees an hour

we know there are 60 nautical miles per degree therefore 60 miles x 15 = 900 miles per hour or 15 miles/minute. That is why accurate clocks were so critical to navigation. To accurately fix your position you would measure how far the sun was off the horizon at an exact time of day. A book tells you the exact location of the sun at that exact time (Greenwich meantime, the place in England where the observatories were located when they wrote the book). You use trigonometry to calculate your position, but for every minute your clock is wrong your math is off by fifteen miles.

A clock measures the same units as a compass, just times 15. The math was simple, isolating it from the sailors kept them on board. The French, not liking the English for making things so complicated developed a simpler system. 400 degrees around the world, 100 minutes per degree = 40,000 kilometers around the world, yes kilometers are just 1/40,000 of the earth circumference. Since the planet is covered in water they went further with the system.

1 kilometer = 1,000 meters
1 meter = 100 cm
1 cubic cm is a "cc" = 1 gram of water (the units of weight comes from a volume of water divided by the unit of distance which comes from the circumference of the earth)
1 cc = 1 milliliter = that is why one liter of water weighs exactly one kilo

The bottom line is that the whole metric system, volume, weight and distance comes from dividing the "water planet" into smaller and smaller divisions of it's circumference and just weighing the little pieces. Also remember there is an old English saying, "a pints a pound the world around" referring to the weight of a pint of water. So the same is true of standard units' volumes and weights as well. Just the math is more complicated because of the "software password" required to keep the sailors in check.

As described in this example, by tuning the math such that the common sailor found navigation difficult, the Captains were able to control the crews. Thus reducing the risk of mutiny and encouraging allegiance. This process strategically controlled adoption rates, logistics, value chain, channels to market and even branding. Branding associated with the prestige of being a ship's Captain. Was the branding diluted with the advent of the simplified navigational system developed by the French? Containing in-depth knowledge of cultural differences and respecting their origins enables the development of effective business strategies.

Cultural - Summary

An education plan is designed to address cross-cultural and industry paradigms. People and corporations tend to become experts in their markets and on their technologies. When an emerging technology appears on their radar screen it is often overlooked simply because of a lack of understanding. The individual assigned to the function of strategic business planning for emerging technologies has to incorporate an educational strategy; most often this is performed during initial disclosure meetings. As was the case with Cambrios, Dr. Pakbaz was attempting to explain their biotechnology to individuals who are experts in semiconductor technology who contain comprehensive knowledge of semiconductor processes. To obtain any attention Dr. Pakbaz had to explain the fundamental biotechnical science behind the Cambrios technology and how it relates to semiconductor processing. Cultural differences can be overcome with an educational plan, a plan designed to educate industry or consumers on the value proposition an emerging technology provides. As demonstrated by the origins of the sailing industry, Amway, Cambrios, and HLL; developing an educational plan, which incorporates cultural differences, will enhance business strategies for many emerging technologies and markets.

Rainmakers, Connectors, Mavens, Guanxi

A valuable resource for promoting emerging technologies is through Connectors and/or Mavens. The word Maven comes from the Yiddish, and it means one who accumulates knowledge. A Maven is a person who has information on a lot of different products, technologies, prices or places. This is the person who connects people and has the inside scoop on the marketplace.

Rainmakers are similar to Connectors and Mavens in that they know lots of people. By having a foot in so many different worlds, they have the effect of bringing them all together. Emerging technologies need exposure. Collaborators need to be identified and brought online. This can be achieved by either substantial investment used to promote the technology raising its awareness in the industry through trade shows, or exposing the technology to a Rainmaker. Implementing both methods has proven to decrease the rate of adoption. A technology based Rainmaker will know which complimentary technologies are required to bring an emerging technology to market and which company is best suited to make that happen. The Rainmaker will maintain relationships with key decision makers within an extensive number of organizations and technological industries. The role of a Rainmaker is one of the most undervalued positions in any industry. Instead of attempting to obtain awareness through publications or opening a booth at a trade show designed to: a) Hope the right individual will walk by, b) The Individual will contain with the much-needed complimentary technology and, c) They will immediately understand the opportunity. Place it in the hands of a Rainmaker and if they are excited about the technology, they will arrange the key meetings and introductions.

> *"It isn't just the case that the closer someone is to a connector, the more powerful or wealthier or the more opportunities he or she gets. It's also the case that the closer an idea or product comes to a connector, the more power and opportunity it has as well."* – Malcolm Gladwell, "The Tipping Point".

Frequently referred to as *Knowledge Networks,* corporations build databases designed to map technologies, markets, companies and contacts to their corporation. The net effect is equivalent to a Connector. An emerging technology requires strategic collaborations: Knowledge Networks are used to identify which sector and company is best positioned for collaboration with the emerging technology.

In China this knowledge network is aided by *Guanxi*. Guanxi is an intricate weave of relationships, which underlies the function of its society. It is believed that a person with strong Guanxi will be very effective in business. Business cards are collected and revered as very important assets for creating Knowledge Networks designed to map key contacts within industry sectors.

Wharton on Managing Emerging Technologies makes a couple important and not to be undervalued statements:

> Networks play a central role in the management of emerging technologies.
>
> ...because emerging technologies are built upon emerging knowledge, *knowledge networks* are central.

Placing emphasis on the significance of knowledge networks, Wharton brings to the industry awareness of the value of knowledge networks. Creating a knowledge network is not sufficient; the network must be maintained to be effective. Companies change business directions, decision makers change companies and sometimes industries. As an example, Exhibit 5.1 demonstrates the effect time had on the telecommunications industry from March 2000 to March 2004. The size of the telecommunications industry in the USA from March 2000 to March 2004 dropped from US$865.6 billion to US$255.3 Billion. Companies such as Worldcom, Cable & Wireless, Northpoint and Genuity filed bankruptcy, while AT&T Broadband was acquired. Key decision makers within these organizations changed companies, technology sectors and geographical regions. Between 2000 and 2004, maintaining relationships within the telecommunications industry would have been very difficult.

Maintaining a database of which company is still in business, which has market share, what alliances exist between companies, what technologies are in R&D, which company is open to collaboration and is best positioned for collaboration is imperative and critical to the success of an emerging technology. One can only imagine the numerous emerging technologies, which were shelved as a result of the telecommunications industry decline.

Technology trends and standards have a similar relationship with time. Standing still or not maintaining contacts and relationships can rapidly result in a loss of perspective and a dilution of a knowledge network's value.

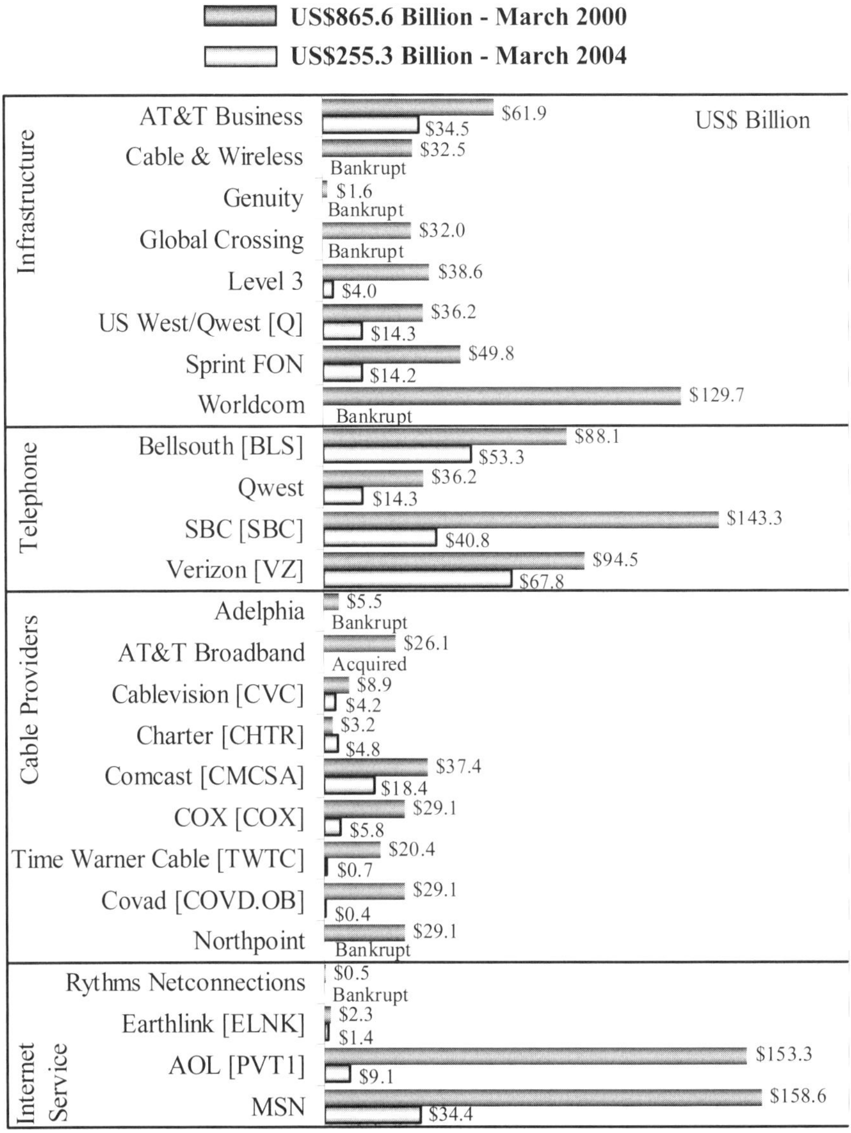

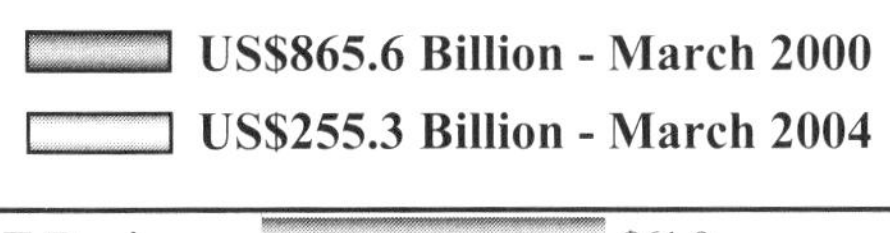

EXHIBIT 5.1 Time effect on Telecommunication Knowledge Network

Source: SEC Filings March 2000 through March 2004

As shown in Exhibit 5.2, a Rainmaker or Connector is in the midst of the network, when exposed to an emerging technology they will analyze the network quickly identifying which path will yield the highest probability of successful alliance or interest. The connector will progress through market or technology sectors and sub sectors, then to companies and the appropriate contact within that company or organization. As an example, suppose a Connector is approached with a novel emissive display technology for illuminating a pixel. The first technology sector is displays. The sub sectors define which type of display technology such as Liquid Crystal Displays (LCD), Plasma, Projection, Poly-LED, etc. Further understanding of the fundamental science behind the novel technology will allow the Connector to identify the sub sector and which companies are investing in complimentary research.

Market share is also critical while evaluating potential collaborators. Market leaders have the ability to define industry standards and rapidly implement an emerging technology. Second and third tier organizations tend to want first tier position and can prove to be more aggressive with adoption. Knowing which organization is best to approach is greatly aided by maintaining personal relationships throughout the industry.

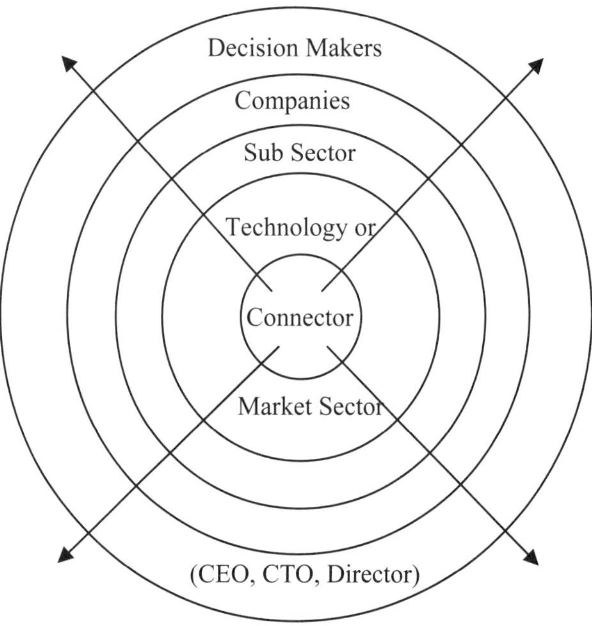

EXHIBIT 5.2 Rainmaker/Connector Network Sectors

Can the connector function be taught, supplemented or orchestrated? One such example is Intel Capital. Intel Capital invests in numerous opportunities focused around industry growth and alignment with corporate strategy. Every year Intel Capital hosts a CEO Summit for their portfolio companies designed to provide networking opportunities between the portfolio companies and industry executives.

In 2005 the Intel Capital CEO Summit was held in Beijing, China. Acknowledging the importance and value of contacts within a technology sector, CEO's from Intel Capital's portfolio companies, along with key industry executives are invited to attend the conference, which is focused around networking. Attendees register online and provide interest in meeting executives and decision makers from various market and/or technology sectors. Attendees receive a list of potentially compatible executives and rate them according to preferred introduction. During the conference there are numerous matchmaking events, which are designed to provide introductions between executives. One such matchmaking exercise is similar to the speed dating process. This is a process where a CEO from a portfolio company sits across from an industry representative and is given 15 minutes to run through an elevator speech in an attempt to determine if there is enough interest to warrant further communication. Within an hour, portfolio executives have the ability to make strategic introductions with several key executives, markets and Intel decision makers. Instead of expecting the CEO of their portfolio companies to have the connector type of personality or ability, Intel Capital provides an environment designed to accelerate the process. The summit included group tours to companies in Beijing such as Nokia Capital Telecommunications Ltd., Lenovo (at the time, the 3rd largest PC manufacturer in the world), Raycom Laboratories, and Nova Electronics Mall (at the time, the largest IT mall in China). The tours provided portfolio executives an exposure to the emerging China market and the requirements to be successful in the market. In addition to the CEO Summit, Intel Capital holds what they refer to as ITD's (Intel Capital Technology Days). This is a process where portfolio managers generate a list of companies, whom they believe have complimentary technology to a tier one organization within the market. The tier one company reviews the list and identifies which companies they wish to meet with and a day is selected in which the portfolio companies provide a limited presentation, no NDA is required. If the industry leading organization wishes to pursue further investigation of the portfolio company's technology, it is up to the portfolio company to follow up and orchestrate NDAs and any other subsequent agreements. Intel Capital not only provides funding, they also provide the highly valued connector function designed to increase the opportunities of success for their portfolio companies.

A clarified perspective on the value a Connector brings to an organization is time and money (rate of adoption). Without a Connector, the emerging technology must spend money over several years (typically 3 to 5) presenting at conferences and exhibiting at conventions hoping the key person will walk by and take notice. A Connector will simply introduce the technology within the week, saving both the investors and the company significant time and money.

Elevator Pitch

Prior to communicating with any potential customer or investor, build an elevator pitch. With the knowledge obtained from the previous chapters, and implementing the NABC value proposition as defined in Carlson and Wilmot's publication, "Innovation: The Five Disciplines for Creating What Customers Want", an effective elevator speech can be created. NABC is the core to the elevator speech, which focuses on the customer's requirements. NABC is represented by four fundamental questions:

1) What is the quantitative customer and market ***Need***?
2) What is the specific ***Approach*** to satisfying that need?
3) What are the quantitative ***Benefits per costs*** from that approach?
4) Why are these benefits per costs superior to the ***Competition and/or alternatives***, by name?

Wrapping this core analysis beginning with a *hook* and ending it with a *close* creates the elevator speech. A hook is a statement or question presented to the audience designed to extract a positive response, a confirmation, represented with a verbal "Yes" or a nod of their head. As an example, a door-to-door solar hot water heater salesman in Phoenix Arizona rings the door bell. When the homeowner answers the door the hook is presented. "Are your heating bills too high?" With a reply of a "Yes", the remainder of the elevator speech is continued. This is a very effective process frequently used all around the globe.

The close of the elevator speech is making the request. It needs to be clear and precise. Whether one is attempting to sell a product, obtain funding, or create collaboration, clearly define the request. Refine the elevator speech; practice it many times to get it right. Work with the team to analyze possible questions, which might be asked and answers to those questions.

While discussing the emerging technology with potential customers, listen to them. They will define what their requirements are. The elevator pitch can subsequently be refined.

PR & Promotion

Informing the industry about the existence of the emerging technology is crucial. If the industry is not aware of the technology, it is a guarantee that it will not be adopted. Care must be taken to target the audience. Attending and presenting at industry conferences which are heavily based upon technology do not typically have customers attending. Possible collaborators can be found at these conferences. It is good to promote the technology through technical conferences but they should not be the major focus. A better solution is to attend conferences at which potential customers are attending. It is at these conferences where one can build strategic relationships and identify specific customer requirements used to better position the emerging technology. An emerging technology typically does not have the budget to build a channel. It is better to look for an entry point into an existing technology channel. Large organizations with existing channels focus on and leverage this with their emerging technologies. Attending conferences where the customers are located can provide this information.

Another route to promote the technology is through technical publications. Write articles for publication in technical journals. Occasionally, resistance to publish occurs. This can be overcome by having another source publish the article or technically related articles. For example, meet with a well known industry analyst to discuss the technology and its opportunities or value added differentiations and have them publish a positive article. Ghost writers can also serve useful in this area. With the expansion of communication through the internet, Blogs are another source to promote and obtain industry awareness of the technology.

Investors and Connectors are heavily networked and tend to focus on technologies along the value chain. Having a good connector on the advisory board can prove to be instrumental in promoting the technology. This is very effective when seeking adoption across different countries.

Letting the industry know that the technology exists and its value proposition requires extensive international travel attending many conferences and visiting numerous companies. This requires a budget with minimal staff and sound planning.

Collaboration Targeting

Gathered market data will reveal industry leaders and followers. Time and resources are to be allocated for a development project with a company who falls in the higher market share category. Lower tier organizations should only receive time and resources if their technology is compelling enough to increase one's value proposition.

Market sector companies can be analyzed to determine which is most likely motivated to adopt an emerging technology. There are two factors which inspire a firm to adopt an emerging technology; motivation and capability of execution. Often an emerging technology will require significant R&D resources to demonstrate proof of concept and processing limitations. As some corporations dilute R&D budgets to the concept that technology can be acquired, they may not possess the resources required to perform the required research. Motivation could be there, but the capability may not.

An effective tool to analyze a firm's motivation is the BCG growth-share matrix. Developed in the early 1970s it positions a firm's businesses across two dimensions, relative market share and market growth rate, Exhibit 5.3.

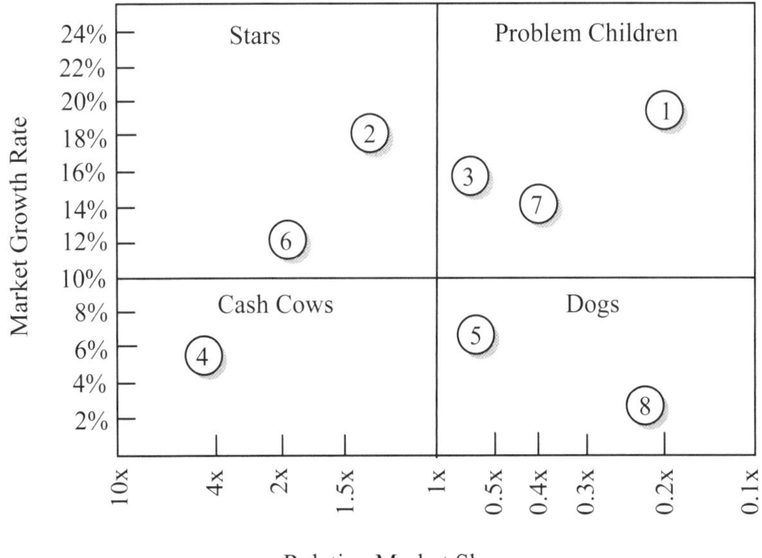

EXHIBIT 5.3 Growth-Share Matrix

Source: Boston Consulting Group

The horizontal dimension is Relative Market Share of the SBU to share of largest competitor. It is fixed and does not change. It is plotted on log scale so that the midpoint is 1.0, the point which the SBU's market share is exactly equal to that of its largest competitor. Values on the Market Growth Rate axis change to reflect actual market conditions.

Products or SBUs which fall within the *Stars* quadrant is a market leader positioned with high growth. *Cash Cows* have high market shares performing in low growth markets. These products or SBUs should be highly profitable due to the high market share and low investment requirements. Revenues from Cash Cows should be used to support the firm's developments in high-growth markets. *Problem Children* are business units or products with low market share in high-growth markets. Firms are in a position to decide whether to invest heavily in an attempt to convert it into a Star performer or divest through end-of-life or sale. *Dogs* have low market share in low-growth markets. They are generally unprofitable and require constant investment to maintain their position. The general recommendation is to divest from these investments.

Analyzing an industry leader's growth-share could reveal they only have offerings in the Problem Children or Dogs quadrants, demonstrating a highly motivated firm positioned for collaboration. When approaching potential collaborators, the value proposition needs to be clearly articulated. Sufficient information has to be disclosed allowing the collaborator to determine if there is interest in following up. Many startup organizations attempt to provide comprehensive presentations which often result in clouding the message. Appendix B is an example of a preferred one page emerging technology primer. It contains pertinent information a collaborator will want to see. It is designed to allow quick assessment and provide the opportunity to invoke questions, leading to follow up meetings.

External Communication

As time is spent developing relationships and promoting the technology the organization needs to present a consistent message to the industry. As employees attend tradeshows and stand in the booth, they will be asked numerous questions from industry representatives. These questions can be role played in advance to conceptualize questions and develop consistent responses. These responses are to be practiced by all members who communicate to potential customers. As industries are heavily networked, representatives will communicate with one another discussing the emerging technology. A common question asked is, "What do you think about the technology and company?" Presenting a consistent message helps build industry trust and interest.

Customer Discovery Process

A book on this author's short list of must reads for one who undertakes the role of managing an emerging technology or associated startup is, "The Four Steps to the Epiphany" by Steven Blank. In his book, he outlines a process designed to align an organization's offering with market expectations/needs. Most product development processes start with a Marketing Requirements Document which feeds into Product

Development. At the end of the product development cycle is "lessons Learned". Here the organization reviews historical data of the project to refine the process. This works well for established organizations and markets with clearly defined requirements. However, startups managing an emerging technology with a target market segment identified require a different approach. Market adoption is achieved by adhering to an iterative process designed to identify the early adopting customer and aligning the technology specifications to that customer. Blank provides a process for startups that focuses on reaching a deep understanding of customers, their problems and discovering a repeatable road map of how they buy, Exhibit 5.4.

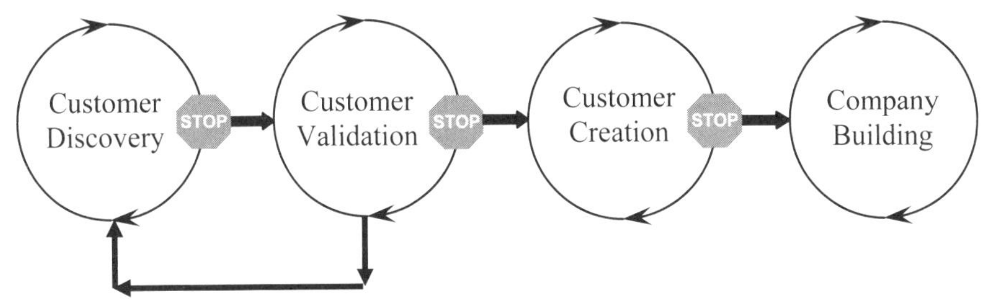

EXHIBIT 5.4 The Customer Development Model

Source: Blank, Steven Gary (2006) *"The Four Steps to the Epiphany, 3rd Edition"*, Cafepress.com

"Empire Building" or Premature Scaling as described by Blank is the immediate cause of the death spiral. Getting the product launch wrong and building an infrastructure behind it, has more often proven to be the death of many startups. Prior to building an infrastructure to launch a product, the market and customers must be fully understood. Blank makes it clear that the Customer Development Cycle is not a replacement for a product development cycle. The Customer Development Cycle includes methods on how to locate and communicate to potential customers. It identifies what questions to ask and what to do with the responses. It does not focus on the "Killer Application" but rather helps identify early adopters. Which organizations are visionary and which are followers. Searching for and identifying which customer not only knows they have a problem but have attempted to create their own solution and budgeted for a solution. These early adopters generate revenue for a startup. Blank doesn't stop there, his process continues on with how to transcend an organization's offering to mainstream market adoption and the organizational changes required while bridging the *Chasm*. Within each phase there exists a circular loop of refinement, designating the process is iterative. It utilizes lessoned learned *in process* rather than waiting until the end for a wrap up review. In this methodology, one keeps cycling through each step until enough success is generated to carry one to the next step. Notice the circle labeled 'Customer Validation' in the diagram has an additional iterative loop returning to 'Customer Discovery'. Customer Validation is a critical check point to determine if you're offering is what customers want to buy.

In each of the steps, the product development and customer development teams meet in a series of formal "synchronization" meetings. If the two groups are not in agreement Customer Development does not move forward. It is the goal of a startup to identify alignment between their offering and the customer who will buy the offering.

Cambrios started out utilizing protein affinity to perform nanostructure fabrications. During the Customer Development cycle, Cambrios' product development team discovered a process where they could build the structures without utilization of proteins, at a lower cost. They abandoned the protein process. By following a similar Customer Development process specific structures needed by the market where Identified. Cambrios' first product offering was an Indium Tin Oxide (ITO) replacement for fabrication of touch screens, utilizing carbon Nano fibers doped with a transparent conductive polymer. Focus was on discovering customers, validation and customer creation. Organizational building of personnel such as Sales and Marketing did not occur until they knew what they were building, for whom and the value it brought to the market.

Gap Analysis

As the Customer Development model is executed, the market will reveal requirements for adoption into specific applications. Communicating these requirements back to the Customer Validation team is critical. There is often a gap existing between the performance specifications of the emerging technology and the specifications an application requires. This can be communicated by the use of a gap analysis presented as a table or spider chart. Exhibit 5.5 identifies performance specifications for Active Matrix Poly-OLED (AMOLED) technology and compares it to the market accepted technology of Active Matrix Liquid Crystal Displays (AMLCD) for the years 2001 and 2006.

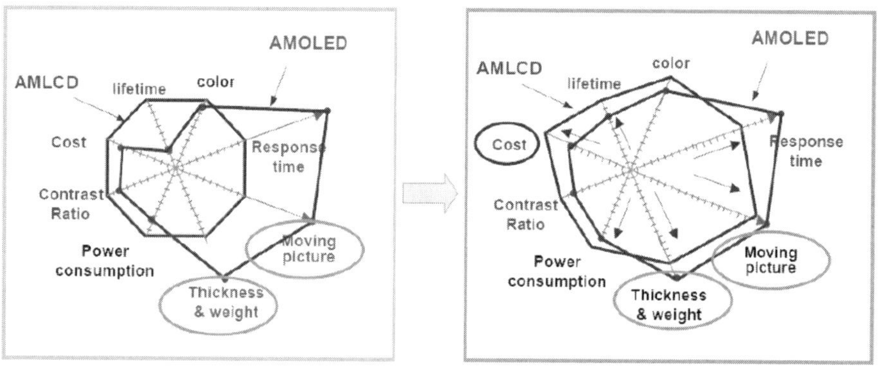

AMOLED vs. AMLCD 2001 AMOLED vs. AMLCD 2006

EXHIBIT 5.5 AMOLED vs. AMLCD Gap Analysis (Spider chart)

Source: Cambridge Display Technologies, Oct 2007

One can observe that a spider chart provides a graphical representation of the data which can be quickly and easily understood. An additional significance is that specifications change with time. As the market continues to improve technologies and processes, specifications will improve. A spider chart in one year will very likely be different the following year, thus providing a moving target for an emerging technology.

In order to displace a currently adopted technology, an emerging technology must not meet the requirements, but rather it must exceed them in performance. Displacement requires additional investment by the adopters and the emerging technology's value proposition must justify such an investment.

Managing Stakeholders

Another critical and ongoing piece of the equation/process of managing emerging technologies is the efforts and patience required to manage stakeholders. When managing emerging technologies, most stakeholders are not aware they are stakeholders. Stakeholders can range from inventors to customers. This includes everyone along the value chain including integrators and material suppliers. Managing stakeholders can be the most tasking part of the process required to bring an emerging technology to market. Under estimating the significance of stakeholder management can be devastating. Several things can and will go wrong. Suppliers can lose interest and quit producing or even collaborate on blocking strategies, investors can loose confidence, and intellectual property can get out of control creating barriers to market and potential legal issues. The best practice for managing stakeholders is communication, structured, consistent communication. Everyone must know what role they are playing and the direction they are going.

Summary

Frequently termed "Managing Innovation", the in-depth knowledge and perseverance required to successfully manage an emerging technology to the point of industry adoption requires a comprehensive understanding of multiple aspects. It will require skills developed around understanding and respecting cultural differences, technical differentiations, relationship management, and strong negotiation skills.

A significant number of business strategies are based upon data analysis alone. As demonstrated, a more comprehensive approach includes aspects of culture. Ongoing communications enhances success. Mavens or Connectors can prove to be effective with the execution of business strategies. Professionals who perform business development are more effective if they possess knowledge beyond the corporate strategy; for they are expected to deliver on the strategy.

Defining one's product or value proposition around the customer's requirements will increase the adoption rate and reduce the time to market. Too often organizations get caught up on performing research and development on interesting stuff. While managing an emerging technology, *ALL* research and development is to be aligned with customer defined requirements. Any business will fail if it does not obtain a customer, and if the business is not listening to the customer, one will not be obtained. Adhering to the Customer Development Model can facilitate customer discovery and alignment.

Some emerging technologies will have milestones to over come prior to full adoption, which can require collaborators with strong market strength. Defining the target product could take significant time and management of industry perception. Managing industry interest and perception requires attendance and participation at technically based and customer focused industry conferences. Perseverance is required when building relationships with collaborators to define an acceptable product specification while increasing over-all industry interest (i.e. managing the hype). Carefully managing agreements with collaborators can have the effect of increasing one's research capabilities. This is achieved by utilizing a collaborator's R&D resources to develop the technology.

Once potential collaborators and value propositions have been defined the corporate message to the market is to be consistent. Developing an elevator pitch coupled with PR planning sends a consistent message to the industry. Establishment of collaborations demonstrates achievement and validation of strategic direction. It provides confidence to stakeholders by initiating industry interest. Discussions with the industry through conferences or direct corporate communication will identify additional applications and help refine the value proposition.

Cultural issues can become apparent early. Communications, terminology, and application issues, if they exist, will begin to appear and will need to be managed. In the Cambrios example, initial communications (choice of words) had to be altered to adjust for cultural differences in technical vocabulary and improve comprehension.

Chapter 6

Legal Considerations

"Change is the law of life. And those who look only to the past or present are certain to miss the future." - John F. Kennedy

This chapter is designed to provide a fundamental understanding of the common types of agreements required. It only provides basic insight and awareness into some of the legal documents and issues encountered on the path to market adoption. It is not designed to be a manual for understanding all the legal requirements and cannot serve as a substitute for experienced legal advice. It only brings attention to some aspects for consideration pertaining to business development.

In the old English days, there were the ecclesiastical courts and the King's court. The former was more geared for substance over form (thus, the courts of equity), whereas the latter was exactly the opposite (a litigant would often lose if he did not follow the procedures to the letter), thus, the courts of law. This distinction of courts has pretty much disappeared today, but the remedies still remain: *equitable remedies* are such things as restraining orders (court orders saying one shall refrain from doing this or that) or specific performance (a court order saying one shall affirmatively do this or that); and *legal remedies* are monetary damages. A discloser who discovers that the recipient of confidential information has improperly disclosed confidential information would want to do two things: (i) get a court order saying that the recipient must stop the disclosure; and (ii) recover money damages for the damage caused by such improper disclosure.

Business strategies require legal protection for market control and corporate value from an international perspective. Whether the strategy is regionally focused or internationally focused the emerging technology will execute numerous contract agreements. As an emerging technology progresses from the laboratory to market, the path is filled with IP applications, Nondisclosure agreements, collaboration agreements, supplier agreements and host of other contracts. Legal representation not only includes corporate relationship protection but also navigation through government legislations and international trade agreements. Established International legal firms typically have specialists on staff to manage the varying sectors of representation. The firms can be leveraged to support in-house or local council.

Legislative Review

A review of legislation (including regulations) for each market region is required to bring a technology to market. Laws and regulations can either inhibit or promote market adoption. As an example, the FDA mandate to use RFID promotes technology adoption. While regulations eliminating the use of full motion video for signage applications in automotive traffic locations inhibits technology adoption.

Regional markets can contain protective legislation inhibiting market penetration. They may also have regulations and legislative mandates which could increase market adoption. To overcome inhibitive regulations, one must decide whether to attempt changing the regulations or implementing strategies with enough market strength to alter such regulations. An example would be Intel's Sean Maloney's collaboration strategy resulting in the adoption of WiMAX. Faced with US$1100 for landscaping costs alone just to install optical fiber to each home, criticism from tech analysts, and resistance from the world's standard setting body the IEEE (the Institute of Electrical & Electronic Engineers). Maloney changed direction from optical to wireless, built strong international collaborations, and re-wrote the rules of wireless communications. His collaboration strategy resulted in the adoption of WiMAX by IEEE and infrastructure building around the globe.

Intellectual Property

As previously mentioned, an emerging technology commonly begins at the material level. This may be the core Intellectual Property. Whether to patent an invention as IP or maintain it internally as proprietary information (such as a trade secret) is a critical decision and should not be taken lightly. Generally an invention is likely to be patented if it is a revolutionary concept. On the other hand, if it is comprised of refined know-how, may be better kept as proprietary information.

Global patent management has always played a role in business strategies. Today's economy, more than ever before, requires strategic thought and execution to protect and obtain maximum patent value. Issues relating to corporate, cultural, and international trends have to be considered. Today's economy has become undeniably international with manufacturers in one country providing components assembled in a second, and marketed in a third. Also a patent can be leveraged beyond the country of invention. It should be patented where it is likely to be manufactured and where it is marketed, but not necessarily in countries of little strategic or economic value in relation to the patented invention.

A patent excludes others from exercising certain rights with respect to the claimed invention:

The right to make
The right to use
The right to offer to sell and sell
The right to import

Each of these rights is to be considered when developing the IP strategy for each country.

The path to market requires the development of material processing, manufacturing processes, applications and business designs. Exhibit 6.1 provides a graphical representation of this path. Each ring represents both an enabling wall and a blocking wall. The potential to reach the market increases as the technology crosses each ring.

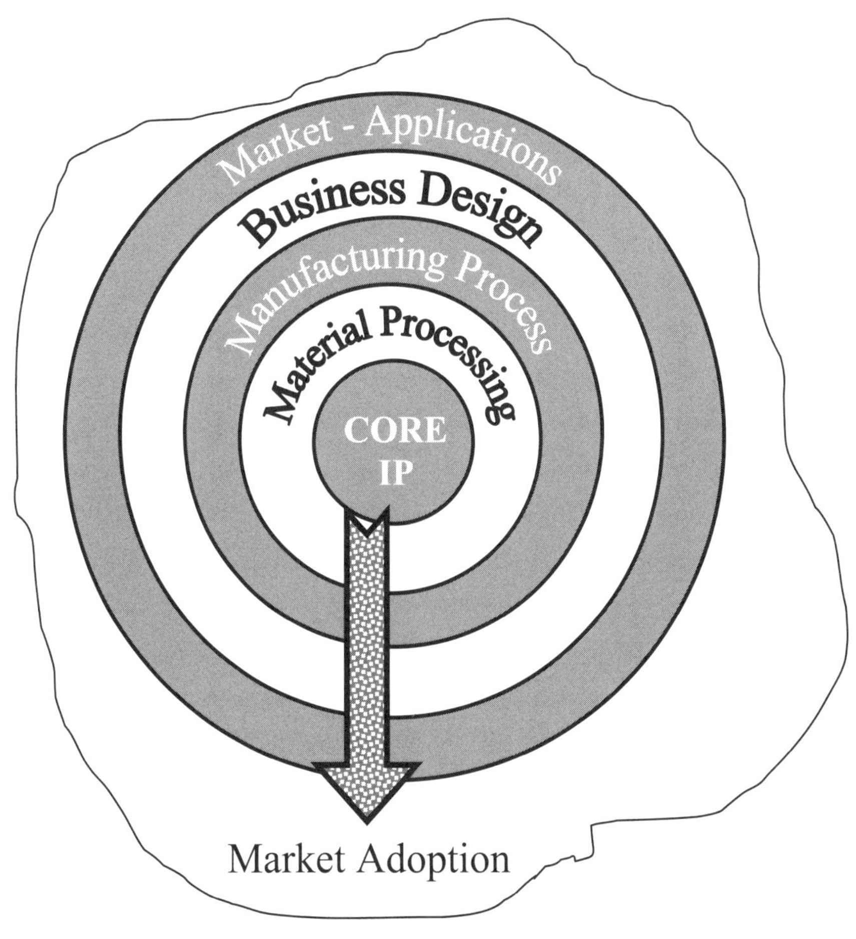

EXHIBIT 6.1 Market Path for Intellectual Property

Both competition and collaboration are critical to successful market adoption, as are access to IP and IP protection and exploitation strategies. IP can serve as an enabler when a company controls the licenses, which provides control of the market share and pricing. Limiting the number of licensees can prevent a flood of product into the market resulting in the technology becoming a commodity. However, IP can become an inhibitor when a single company attempts to control ALL patentable aspects of a technology. As competitors are squeezed out of the market, it may result in market rejection instead of adoption.

A company may encounter a block on the path to market when a related invention has been patented by a competitor. The solution may lie in securing a license of the related invention from the competitor; design around the related invention, or both. Accordingly, it is critical to identify and establish relationships with strategic collaborators, so as to compliment the path to market.

Blocking is exampled with the "Stool and Chair invention concept" which is familiar to virtually all patent practitioners.

> A portable seating appliance art *stool* having a platform with three legs was invented by Mori. He applied for and received a patent for his invention.
>
> Another inventor, Chang, came up with an invention *chair*, in which he added a fourth leg, a back support and arm support to improve comfort. The chair is an improvement over the stool, thus the patent office granted a patent to Chang for his invention.
>
> The interesting question is: Who can do what and who can exclude whom?

Mori invented the basic portable appliance art *stool*. He can exclude others from making, using or selling his *stool* invention. He can also make, use and sell his stool invention without any fear of infringing upon other's patents. He has both negative and positive rights on his invention.

On the other hand, Chang can only exclude (negative right) others from making, selling or using devices which are comprised of four legs, a back support and arm support. His patent would not extend to any right to make those devices without infringing third party rights. This is because *chair* is also comprised of a platform and three legs (which is equivalent to *stool*), if chairs exploited without obtaining an appropriate license from Mori, then such activity would be infringing Mori's stool patent.

This example demonstrates that a person having a patent can be in a position where they have an inability to fully exploit their invention. In other words, a fundamental patent has both positive and negative rights, whereas an improvement patent would only enable its owner to exclude others from exploiting the patented improvement.

Business perspective of IP

As access to IP and IP protection and exploitation strategies are fundamental to maintaining market share and executing business strategies, tough questions need to be asked such as:

1. Has the intellectual property for the technology been received? Applied for?
2. How many patents filed and for how many applications?

3. Does the intellectual property encompass enabling technologies and processes?
4. Does it rely on complimentary intellectual property owned by another organization?
5. If another company or organization controls the IP, does IP opportunity exist for the controlling zones to market as described in Exhibit 6.1?
6. What is the controlling company's business model or licensing model?
7. What does the intellectual property (IP) map look like?
8. Which patents are critical?
9. Are patents properly exploited throughout the supply chain?

Patent searches and review of the patent portfolio provide some comfort that there is a clear path to market. A thorough patent search, or disclosure from the market, may reveal patents which may block the path to market.

Additional decisions on IP strategy are to be discussed. For example, Taiwan is not a member of any international patent treaty. What is the strategy for Taiwan? Should patents be filed in Taiwan? Everywhere except Taiwan? Sell or license Know-How to a targeted organization within Taiwan? The market is global and all strategic planning efforts are to entail a global perspective.

Intellectual property has significant value for not only the corporation, but for the regional economy as well. As depicted in Exhibit 6.2 research and development of IP demonstrates a linear correlation between a country's GDP and the number of patents granted in that country. The three largest global economies, namely the United States, Japan and Germany, have significant programs dedicated to fund technology research and development. This demonstrates the value of IP in providing a competitive advantage.

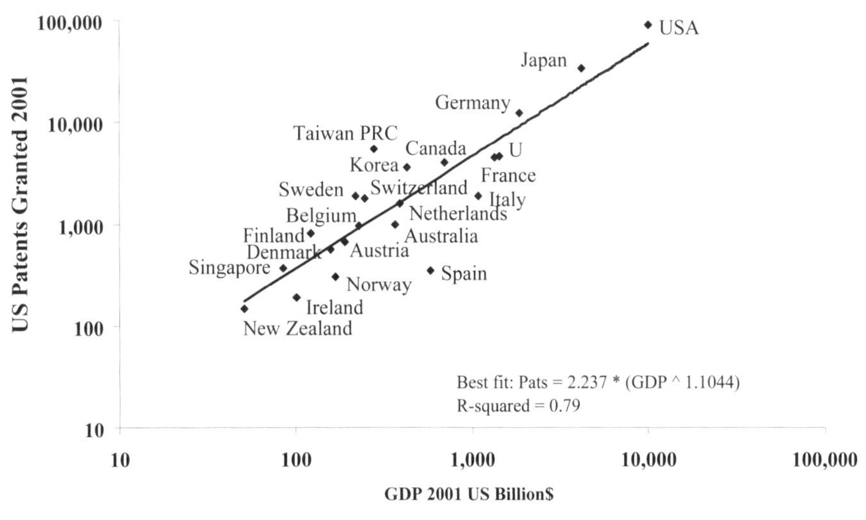

EXHIBIT 6.2 2001 Granted Patents vs. Country

Source: Anthony Breitzman. Reprinted with Permission.

IP Ownership Strategies

Startup organizations like the concept of granting non-exclusive licenses to everyone. This approach provides a perception of maximum revenue potential. However, this approach can be out of alignment with market leaders, who may desire an exclusive license for strategic market control to protect their R&D investments. Both organizations will negotiate from their business objectives. Often a compromise is reached where the licensee will receive a two-year exclusive license, with certain milestones. Milestones are outlined to ensure the licensee is progressively exploiting the technology to market and not prohibiting growth of the licensor.

Licensing to a tier 1 organization who has the channels to market brings substantial value to the organization. Whereas licensing to a lower tier organization that does not have channels to market, brings little value to the organization. As described in chapter 4 a critical aspect of any business strategy is strategic control. When negotiating for non-exclusive or exclusive, consider the channels to market the licensee brings.

In 2001 and 2002, the display market for mobile phones was dominated by three companies, Epson, Samsung SDI, and Philips (Appendix Table A.8). From a material perspective (number of units) these three companies represented more than 65% of the market. Data for this market sector indicated thirteen other companies had market share worth mentioning which did not fall into the "Others" category. One could track which of the three dominant players were increasing market share and which were loosing market share. This information is valuable while negotiating an exclusive license with a market leader. For if one obtained a design win allocated to 30% of the products manufactured by Epson (9366 units), they could quickly obtain 7% market share. In this scenario, granting a two-year exclusive license would allow a startup organization time to focus on building capacity, and provide first mover advantage for the licensee. This scenario brings rapid growth and significant value to stakeholders of the emerging technology.

NDA (Non-Disclosure Agreements)

Disclosure of the technology requires the execution of an NDA. No matter how simple an NDA appears, it needs to be reviewed for content and structure. The most common type is a mutual NDA, which protects disclosures of information between two companies. Sometimes multiple company Non-Disclosure Agreements are signed due to the necessity of complementary technologies required to move the emerging technology from the laboratory to application. Some companies adopt a two level system, which is comprised of a CNDA (Corporate Non-Disclosure Agreement) and a Supplement Disclosure NDA. The CNDA covers an accepted mutual Non-Disclosure policy between organizations from a high level perspective, while the supporting Supplement Disclosure NDA specifies the item of interest, terms, and duration.

It is customary to use the NDA of the company disclosing the information. As an example of why it is important to review an NDA for content, the following text has been observed under "Right to Use".

The Confidential Information disclosed may only be used by the discloser, for the benefit of the recipient. The recipient shall have the unrestricted right to use the Confidential Information of the discloser, subject only to any patent or copyright of the discloser.

In this example the receiving company can use confidential information disclosing company disclosed under the NDA without restriction. If a company desires to restrict the use of its confidential information to be disclosed under the NDA, such a company should change the wording. To achieve such a goal, a mutual NDA might incorporate the following:

The recipient shall not make any use whatsoever at any time of the Confidential Information disclosed by the Discloser except (i) to evaluate internally whether to enter into the currently contemplated business relationship with the Discloser, and (ii) such other purpose explicitly set forth in a written Supplement for Disclosure signed by both parties. The Recipient shall, at all times, use the Confidential Information disclosed by the Discloser for the benefit of the Discloser.

Occasionally when such contents are identified and a request for rewording is submitted, it can be viewed as a challenge between legal departments, resulting in a degradation of the relationship and a lost opportunity. When a relationship degrades, both stakeholders have lost. Managing stakeholders is not an easy task. Relationships are intertwined with cultural, professional, communication, and sometimes-emotional issues. An individual attempting to manage the process of strategic business planning for emerging technologies cannot under estimate the task required to manage stakeholders. In the early stages, most stakeholders are not aware they are stakeholders, thus may not posses the ability to apply value or the desire to collaborate.

Letters of Intent (or Memoranda of Understanding)

A letter of intent (LOI) (or memorandum of understanding [MOU]) may be used to lay down the key terms of a deal. For example, two companies may have worked out the main terms of a collaboration, reflected in an LOI or MOU, which they expect to be integrated into a final expression of a collaboration agreement.

LOIs or MOUs can be of a binding or non-binding nature. If of a binding nature, the companies must integrate into the final agreement those terms as set forth in the binding LOI or MOU; otherwise, the company refusing to do so may be facing specific performance and/or damages. On the other hand, if the LOI or MOU is of a non-binding nature (which should be made clear), a party should be able to change the terms of the original deal or even walk away from it without incurring liability. This may, however, significantly impact the credibility of the company seeking to change terms or walking away from the deal entirely.

Collaboration Agreements

Many companies form collaborations for the purpose of developing or exploiting their technology. In addition to the existing technology and related IP rights, which each party brings to the table, new technology will arise as a result of the collaboration. Ideally, IP rights that arise as a result of the collaboration are apportioned between the parties in a manner that permits each party to pursue its ultimate (post-collaboration) business goals. Some basic questions for establishing collaborations are:

1) IP ownership formula?
2) Sole and joint IP ownership definitions?
3) Governing law (US law)?
4) Who files and prosecutes individual and joint patents?
5) Agreement to inform on filing patent applications?
6) Who defends alleged infringements?
7) Who enforces patent rights (joint and separate)?

A number of schemes exist for apportioning IP rights between collaboration partners. The simplest schemes involve designating all collaboration IP as joint IP, or as the property of one party. However, most collaborations adopt one of two basic formula; an inventorship formula in which IP rights follow inventorship, or a technology formula in which IP rights are allocated based upon technology. Also, collaboration partners may adopt a third approach, essentially a hybrid of the two. The choice depends in large part on the structure and goals of the collaboration and post-collaboration.

Inventorship Formula

Applying an inventorship formula to a collaboration between Company A and Company B inventions made by the employees of Company A are assigned to Company A, inventions made by the employees of Company B are assigned to Company B, and inventions made jointly by employees of both Company A and Company B are assigned jointly to Companies A and B. The assignments are made regardless of the nature of the invention. As example, such an agreement will state that each party will own inventions made by its employees, and inventions made by employees of both parties shall be jointly owned. Inventorship will be determined by US patent law, and disputes regarding inventorship or ownership are submitted to mutually acceptable outside patent counsel. The consent of both parties is required to exploit a joint invention, which sets the stage for co-promotion sharing. Both companies may jointly enforce patents against alleged infringers, and if one company opts out of enforcing such patents, then the other party may do so at its expense and keeping all recoveries. Each company is obligated only to defend itself (and its customers) against third party claims of infringement.

There are several potential problems that must be thought through in advance if the inventorship formula is adopted. One problem that often arises is that the non-inventing party may require access to collaboration IP in order to carry out its further obligations under the collaboration, or to pursue its post-collaboration business. Thus, licenses may be necessary to ensure access to solely owned IP rights of the other party. Another problem is determining inventorship rules apply. For example, inventorship rules differ from country to country, and if the parties reside in different countries, each may want its own country's inventorship rules applied. The parties also must determine who pays for joint patents, especially if patent protection on the joint invention benefits one party more than the other. The parties must also decide whether there should be any limitations on the future exploitation and enforcement of joint patents as well defending infringement claims against the invention claimed by the joint patents, as patents often out last collaborations.

Technology-Based Formula

When applying a technology-based formula to the same collaboration between Company A and Company B; inventions relating to a defined technology or field are assigned to Company A, and inventions relating to a different yet defined technology or field are assigned to Company B. This formula also may designate inventions in certain other technologies or fields (that are neither Company A's or Company B's) to be joint inventions.

Several potential problems exist when this formula is adopted that must also be thought through in advance. For example, determining the allocation of inventions to the different fields is often problematic. This task often resides with patent attorneys, a patent committee or a steering committee. Still, disputes routinely arise about which inventions should be allocated to which field, especially if the fields are not precisely (and presciently) defined. Some inventions may be outside of both parties' defined fields, or be encompassed by both fields.

Hybrid Formula

When applying a hybrid approach in a collaboration, the parties may agree that the inventing party will solely own the invention (or if jointly invented, jointly own the invention) except for certain defined technologies or fields for which the technology-based formula would be applied.

Problems inherent in both the inventorship and technology-based formulae are often apparent in hybrid approaches to IP ownership.

Allocation of IP rights based upon inventorship is commonly favored over technology-based formulae where both parties are committing substantial resources and accepting substantial risks in the development effort. On the other hand, technology-based formulae are favored where the goal is to protect or improve an organization's core technology. The many possible variations in the way allocation of IP ownership can be implemented provide the parties with flexibility to negotiate a scheme that works for both partners in the context of the collaborative relationship.

Know-how or Knowledge Transfer Agreements

Knowledge transfer agreements are similar to IP licensing agreements in that they can be written to provide revenue for defined field of use. Where non-patented technology or information is the subject matter of the knowledge transfer, one of the key benefits of a know-how (including trade secrets) or knowledge transfer agreement is that such technology or information are not subject to the life spans that patented technologies are. Absent a publicity event, end-of-life as to permitted use of non-patented technology or information may be negotiated in the terms of the agreement.

Summary

Business strategies require legal protection for market control and to maximize corporate value. Thus IP and business strategies must be defined not only from a global perspective, but also, from the perspective of each targeted country where market adoption of the technology is likely.

Analysis of its IP portfolio and patent clearance activity to ensure the organization has control of any patents which may otherwise block routes to market are essential. In many instances, collaborations, licensing or purchasing patents may be required.

Getting a technology to market will often require negotiation of a variety of agreements. Although NDAs are often the starting point as far as such agreements are concerns, they should not be taken lightly, but reviewed and evaluated carefully.

There are at least there basic approaches to address IP developed in the course of collaborations.

That said, it is evident that highly qualified and experienced legal counsel can prove to be instrumental in developing and protecting market opportunities and strategies.

Chapter 7

Financial

"If performance measures are to create an accurate picture of the company on which we can rely, then the strategy is the tool we use to guide our focus, like operating a camera.... every camera takes a slightly different picture; build one badly, and the output will be shadows and distortion." - Jean Cunningham and Orest Fiume

Funding Cycles

The decision to bring on investors which will provide capital required to move an emerging technology from the laboratory to a successful market is a common consideration. The venture capital market is distraught with horror stories of malicious, manipulating, and controlling behaviors. For example, a technology analyst was describing to an investor, a business model where he discovered an inventor who only wanted to sell the IP and was willing to give 50% to the individual helping him achieve the sell. The venture capitalist commented, "You mean that after you receive the money from selling the IP, you would actually give the inventor his share? You wouldn't keep it? You know he can't afford a legal battle. You can, especially when you are holding all the money". As with every industry, a bad few can destroy the reputation of the majority. The majority of the venture capital community is comprised of highly reputable and skilled individuals. They view investment opportunities as partnerships and want their investments to be successful. They actually need someone to run the business.

There is a difference in funding approaches between large corporations funding an organically grown opportunity and investors funding a portfolio of opportunities. Large corporations who understand the market tend to analyze an opportunity and the required investment to achieve success. Then they will implement their strategy fully funded. Whereas an investor who does not fully understand the market is more risk adverse and will fund based upon milestones. In a situation where an opportunity would require a US$25M investment to achieve a projected ROI in five years, a large corporation will invest the US$25M. Whereas a more risk adverse investor might invest US$2M for one year to hclp achieve a milestone and then invest further as subsequent milestones are achieved. The clear difference between these two approaches is time to market. It is easier for an investor to loose US$2M than US$25M. Thus, the investor will attempt to operate and build the opportunity on a shoe-string budget resulting in a longer time to market. Synergies and momentum are not established and this slow development process can end up without the infrastructure required to achieve success in the market. Additional risks can include not having the ability to recruit highly skilled resources due to insufficient funding. Numerous successful businesses were started with this later approach. For it also prevents empire building and focuses the team on critical milestones.

It is crucial for innovators to understand investors and effectively manage the relationship. Innovators, being the creator of the technology, often want to retain a controlling interest in the startup company. They also expect their exit share of ownership to be equal to their entrance share. However, these approaches can inhibit progress with investors. Innovators need to realize that as projects progress; financial returns vary according to the remaining level of risk.

This next section will attempt to explain the process of investment and dilution from a venture capitalist approach. Exhibit 7.1 shows the sequential funding cycles for a typical startup, along with risk and the associated discount rates typically applied for the level of risk.

Increasing		RISK	Decreasing
Stages of Investment	Seed/Startup	1st Stage/2nd Stage	3rd Stage/Bridge
Required IRR	40-90%	30-50%	20-30%

EXHIBIT 7.1 Pre-IPO Stages of Investment for Startups

Seed Funding is used to provide resources for feasibility into a concept. This is more of a research stage with no business plan and very little assurance the technology will work. Seed investors can be friends, relatives, Angel Investors, or Early-Stage investors. An Angel investor or Early-Stage investor will provide basic business advice and assistance with writing of the business plan. If the technology achieves maturity through value added milestones then seed funding can be used to help recruit key management which is often required for startup funding.

Startup Funding requires more significant funds built around a business plan. The startup will be able to demonstrate knowledge of the market, a planned route to market, prototypes for the technology with limited R&D remaining, and demonstrate technical differentiation along with value proposition to the customer.

1st Stage Funding is for ongoing business. A startup is generally not profitable and is not expected to be profitable. However, a plan for revenue is expected to be outlined. First-Stage funding is used to develop the business by establishing initial major marketing efforts, and definition of product requirements as defined by the customer's needs. First-Stage investors are more actively involved in problem solving and monitor head count to ensure staffing levels are matched to the organizations requirements and attainable sales levels. Majority of the funds are applied to refinement of the technology. Additional key resources are also brought in. Risk to investors is lower because the earlier stages funded the development of the initial proof of concept and promotion of the technology.

2nd Stage Funding is used to finance operational requirements. It is typically used for working capital, fixed asset requirements, and to support the initial growth of the organization and sustainable sales. In the advent of liquidation, assets are more easily recovered thus lowering the risk.

Bridge Funding is intended to carry a company until its Initial Public Offering (IPO). It is expected that the company will go public within a year. Funds can be used to cover ongoing capital requirements, or buy-out an early-stage investor who might be eager to liquidate their funds. Bridge Funding is not required prior to an IPO. Bridge investors typically only hold the stock past the IPO date.

Restart financing, commonly known as emergency or sustaining funding is used to bail out a troubled firm. Stock is sold at a price significantly lower than the previous round. The stock price will be low enough to offer a projected high rate of return.

Another method that can be used to generate operational funds for a troubled or overly capitalized firm is executing an IPO into a Microcap. Issuing an IPO into the Microcap market can generate operational capital required to sustain the business or buyout early stage investors.

Other terms commonly used are Series funding. *Series A* refers to early stage funding which includes Seed Funding and Startup Funding. *Series B* funding includes everything after Series A and prior to IPO. *Series C* is the funding received from the IPO. There can be several rounds of funding for each stage of investment.

Pre-money and Post money Valuation

Some early-stage investment negotiations resemble a game of a poker game. Each player withholding information and attempting to convince the opponents that his hand is better than it actually is. But valuation negotiations are not card games. Unlike poker, the objective of investment negotiations is for investors and entrepreneurs to share information as openly and completely as possible, working together toward a common goal of building a successful company.

Pre-money refers to a company's value before it receives outside financing or the latest round of financing. Post-money refers to a company's value after it gets outside funds or its latest capital injection. Pre-money valuation refers to the value of a company not including external funding or the latest round of funding. Post-money valuation, then, includes outside financing or the latest injection. It is important to know which is being referred to as they are critical concepts in valuation.

Suppose you are managing a startup and while discussing the value of the company with an Investor and the company's value is agreed to be US$1,000,000. The investor desires to inject US$250,000. The ownership shares will depend upon if the valuation is pre-money or post money. If the valuation is pre-money, the company is worth US$1,000,000, the value is US$1,000,000 before the investment and changes to US$1,250,000 after the investment. If the valuation is post money, then the company will be worth US$1,000,000 after the US$250,000 investment.

	Pre Money Valuation		Post Money Valuation	
	Value (US$)	Percent (%)	Value (US$)	Percent (%)
Management	1,000,000	80	750,000	75
Investor	250,000	20	250,000	25
Total	1,250,000	100	1,000,000	100

EXHIBIT 7.2 Difference between pre money and post money valuation

As Exhibited in 7.2, the selected valuation method used can affect the ownership percentages in a big way. This is due to the amount of value placed on the company before investment. If a company is valued at $1 million, it is worth more if the valuation is pre-money compared to post-money because the pre-money valuation does not include the $250,000 invested. While this ends up affecting the management's ownership by a small percentage of 5%, it can represent millions of dollars if the company goes public.

This topic gets very important in situations where a startup has a good idea but few assets. In such cases, it's very hard to determine what the company is actually worth and valuation becomes a subject of negotiation between the entrepreneur and the venture capitalist. Achieving an *agreed upon* value of the company is critical and serves as a base line for subsequent investment discussion.

Valuation and Dilution

Most entrepreneurs focus on the exit value of the venture when negotiating investor share, US$2 Million for 30%. Seed and Startup investors understand that a venture will lose money for the first two or three years, and in order for it to grow to US$50 or US$80 Million, the venture will require a substantial infusion of cash. A common understanding of valuation, discount rates, dilution, and revenue will prove to be instrumental in building a constructive relationship between the entrepreneur and the investor. The following example of a simplified arbitrary company (ID-5) steps through the funding process to demonstrate how this works. This example is simplified; studying it will not make anyone an expert. It is designed only to provide a basic understanding. This example will not analyze detailed market data or the collaborations required to make the business model a success; it will only provide enough information for a basic understanding of the numbers used.

A company named "ID-5" comes with an innovator who has a novel concept for a chip-less RFID solution, low-cost, has a 10 meter reading distance and can be applied to the product at very high speeds. To test the concept, capital equipment will need to be acquired and modified, at an estimated cost of US$525,000. If successful the technology will enable the application of RFID solutions at the customer's factory. The value proposition is to provide low cost machines enabling RFID applications in-line with production. The machines will print the chip-less RFID directly on the product level package, program it and then validate it. The business model is to sell machines and materials to product manufacturers. This business model moves away from the concept of creating billions of programmed RFID tags on a roll and shipping them to the customer. It is estimated that the RFID printed cost will be at, or below, US$0.02 per tag.

The first step is to estimate the value of the opportunity. Performing valuation calculations for startups can be very difficult. Most often, comparable data is not readily available. The Coding and Marking market performs a similar function. Most magazines need a mailing label. Printing the mailing label requires the ability to change the text for every magazine. Small affordable machines are sold and placed in production. The revenue model includes machine sales, consumables such as ink and occasionally paper. This business model and market is similar to ID-5's business strategy and can be used as a reasonable comparable. A quick review of the financial numbers from an investor's perspective will provide insight on what to anticipate, establish a baseline for negotiations, and promote constructive investor relations.

Step 1: Gather market data

As previously stated ID-5 will manufacture RFID equipment into a market similar to the coding and marking sector. The Three leading providers in that market space are Danaher, Dover Industries, and Domino Printing. Exhibit 7.3 is a list of the top three coding and marking organizations as determined by JP Morgan. The number of interest is the Price Earnings Ratio (P/E or PER). Identify realistic comparables and use a conservative PER based upon those companies.

Company	**Source**	**P/E Ratio**	**2006 Sales[1]**
Danaher [DHR]	NASDAQ	21.97	$872M
Dover [DOV]	NYSE	17.94	$607M
Domino Printing [DNO]	LSE	19.18	$396M
Industry Average	NASDAQ	18.74	$1,875M
ID-5	Projected	15	- - -

EXHIBIT 7.3 Comparable P/E Ratios, January 2007

Source Data: 1) JP Morgan Securities Report September 22, 2006, Product ID Competitive Financial Data

To determine the potential size of any market, refer to industry sector marketing organizations. One such organization for the RFID market is IDTechEx. According to IDTechEx, the global market for RFID cards and systems will pass the US$3 Billion mark in 2008. The numbers are rough estimates but sufficient enough to demonstrate the potential size of the market to an investor. Business case studies for RFID are not intended for this example.

Step 2: Generate a financial profile

Use realistic, thoroughly analyzed data to estimate the investments, revenue, and timing that will be required to support the business model.

Year	**1**	**2**	**3**	**4**	**5**
Investment ($M)	$0.525	$12	$0	$50	$0
Net Income ($M)	$0	$0	$2.5	$5	$11
Equity (Out Standing)	$0.525	$12.525	$12.525	$62.525	$62.525

EXHIBIT 7.4 Financial Profile for ID-5

Net Income is defined as profits after Interest, Tax, minority interest, and preference dividends (but not ordinary dividends). It represents the amount potentially available to shareholders.

Step 3: Calculate Terminal Value & Required future values for investors

This example will calculate the percentage of investor ownership required for the three stages of funding received by ID-5. Estimates for the required IRR of each investor can be determined from the discount rates identified in Exhibit 6.1 and other factors such as confidence level of management team to execute the business strategy, risk of technical maturity, relationship with the investor and other similar types of quantitative data.

Seed/Angel Investment:	US$0.525 Million
Required IRR (Determined by the investor):	50%
Startup Investment:	US$12 Million
Required IRR (Determined by the investor):	35%
Seed/Angel Investment:	US$50 Million
Required IRR (Determined by the investor):	25%
Term:	5 Years
Year 5 Net Income:	US$11 Million
Year 5 PER: (Exhibit 8.2)	15

Terminal Value is the future value of the company available to investors and management. The first step is to calculate the Terminal Value of the company using the Net Income estimate at year 5 identified in Exhibit 7.3.

Terminal Value = PER x Terminal Net Income
= 15 x US$11 Million
= US$165 Million

Note: There are problems using PER to establish a "Fair" Terminal Value for a company. It overlooks the fact that companies within the same industry can have different capital structures, returns on invested capital, and growth rates. Investors in startup technologies understand this and know that early stage financial projections are truly a "best educated estimate". A carefully designed and comprehensive multiples analysis can provide in-depth insight to a company. During startup funding there are no financials available for a multiples analysis. A multiples analysis will prove to serve useful during late stage funding cycles.

Step 4: Determine Investor Ownership Required

Using the venture capital formula the final ownership required by each investor can be estimated:

Final Ownership Required = Investment/[Terminal Value/$(1+\text{IRR})^{\text{years}}$]

The required Future Value for the investment = $(1+\text{IRR})^{\text{years}}$ x Investment
= $(1+0.50)^5$ x US$0.525M
= US$4M (Angel Investment)

Next, determine the percent ownership the investor will require at terminal year to achieve their financial objectives. Final Ownership for each investment can be calculated using the basic valuation formula:

Final % Owned = [Future Value (Investment)/Terminal Value (Company)] x 100

Angel Investment = [(1.5^5 x $0.525M)/(15 x $11M)] x 100 = 2.4%
2nd Stage = [(1.35^3 x $12M)/(15 x $11M)] x 100 = 17.9%
3rd Stage = [(1.25^2 x $50M)/(15 x $11M)] x 100 = 47.3%

These numbers represent the percent of final ownership required at terminal value. An early stage investor must purchase enough shares to accommodate future dilution as a result of future investment stages. The calculation above does not include risk. The Angel investor will want a significant higher percentage to accommodate the higher risk. For this example the Angel investor would expect 40% to 90% ownership as identified in Exhibit 7.1. To determine the amount of shares an early investor must acquire in anticipation of dilution their Retention percent must be calculated. Percent of Retention is the ratio of the percent held at terminal year to the percent currently held.

Retention % = (Final % Owned)/(% to be Acquired)

Retained ownership can also be viewed as the percent reduced by the dilution of the next investment round. For example an early stage investor can own 60% equity. If the next stage investor acquires 30% of the company, the early stage investor's retained equity will be reduced by 30%. Therefore:

Retention % = [1 – (total of all future % ownerships)] x 100

Since the Angel investor will be diluted by the second and third round investors by 17.9% and 47.3% respectively, he will only retain 1 – (17.9% + 47.3%) of his original investment. The second stage investor will retain 1 – 47.3% of his investment and the third stage investor will retain 100% of his investment. With these equations, determine how much equity each investor must acquire to achieve their terminal ownership requirements.

Acquire % = (Final Ownership %)/(Retention %)

Angel Investment % = 2.4% / [1- (17.9% + 47.3%)] = 6.9%
2nd Stage Investment = 17.9% / (1 – 47.3%) = 34%
3rd Stage Investment = 47.3% / 100% = 47.3%

Calculations for the funding and dilution cycles of ID-5 have been completed for each investment stage and tabulated in Exhibit 7.5.

Funding Round	Seed/Angel	Startup	Stage 1	Stage 2	IPO
Time (Year)	1	2	3	4	5
Investment ($ Million)	0.525	12.0	0	50	0
Net Income ($X million)	0	0	2.5	5	11
Management Equity %	93	61.4	61.4	32.3	32.3
Angel Equity %	7	4.6	4.6	2.4	2.4
2nd Stage Investor Equity %		34	34	17.9	17.9
3rd Stage Investor Equity %				47.3	47.3

EXHIBIT 7.5 Funding and Dilution Cycles for ID-5

These numbers are based upon a terminal value of US$165 Million. Changing the terminal value affects the investor's return on investment. It is imperative to achieve or exceed the financial goals defined and to execute an effective business strategy. When revenue and the resulting net income are not properly managed investors can experience frustration. This reiterates the necessity to properly understand and execute stakeholder management. This example uses a five (5) year financial plan to explain dilution and does not discuss the different types of stock classifications and other important aspects, such as exit rights liquidation rights, etc.

Cash Flow Management

Upon receipt of funding the CEO is expected to manage cash flow and deliver the numbers to shareholders. Cash flow for a startup is to be managed similar to a company undergoing a turn-around process. Financial planning is to be structured around the corporate objectives of creating shareholder value. Share holder value balances the cash flow from operations, the discount rate (market cost of capital), and debt.

Financial planning is comprised of two elements: *objectives* and *strategy*. The most obvious objective is profit; creating shareholder value through dividends and capital gains. Strategy is comprised of tactics on how to achieve the objectives. The most relevant accounting measure for shareholders is Return on Equity (ROE). This represents an investor's most obvious stake with their investment.

$$\text{Return On Equity (ROE)} = \frac{\text{Net Profit}}{\text{Equity}}$$

Return on Assets is significant to the manager since they are expected to provide an acceptable return on *all* funds invested into the business, invested or borrowed.

$$\text{Return On Assets (ROA)} = \frac{\text{Net Profit}}{\text{Total Sales}}$$

Investors and potential acquirers will analyze an emerging company's performance by focusing upon "Value Drivers". Value drivers for startup organizations are comprised of items such as achieving milestones and establishment of collaborations with market leaders. Development of such collaborations demonstrates industry interest in the technology and provides confidence in their investment. From an operational perspective these fundamental value drivers are comprised of component elements derived from the accounting rate of return.

$$\text{Return On Capital Employed (ROCE)} = \frac{\text{Profit}}{\text{Capital Employed}}$$

$$= \frac{\text{Profit}}{\text{Sales}} \text{ X } \frac{\text{Sales}}{\text{Capital Employed}}$$

In his book Marketing Management and Strategy, Peter Doyle analyzed and broke down the financial aspects of these value drivers and mapped them as levers for improving profitability. He used Chemco's financials to demonstrate how these levers can be used to implement the strategy aspect of financial planning. Exhibit 7.6 maps current levers against target levers. Results from decisions upon where to make adjustments can easily be observed. For example, an increase in unit pricing by 5 percent increases profits by £16.5M, if unit volumes remained the same. With an elasticity of 1 (i.e. a 5 percent increase in price leads to a 5 percent decrease in volume), net sales remains unchanged. Adding a reduction of unit costs by 4 percent decreases cost of goods sold by 11 percent, resulting in an increase of gross margin from £95M to £120M. This translates into an increase in profit margin from 3.9 percent to 12.7 percent. Similarly by planning and controlling capital turnover management can improve return on capital and return on equity. For short term effect, inventory (stocks), debtors and cash can be changed relatively easily. Inventory can be reduced by improving forecasting with sales and encouraging suppliers to implement just-in-time delivery systems. Some companies have poorly organized collection processes. Reducing customer time required for payment terms could result in significant savings. Underutilized cash could be a costly investment. Strategic reductions in cash should produce significant economies. The *key* is to analyze the levers strategically. Small changes in one lever could have dramatic effects on the ROCE. Sometimes laying off resources may not be the best decision.

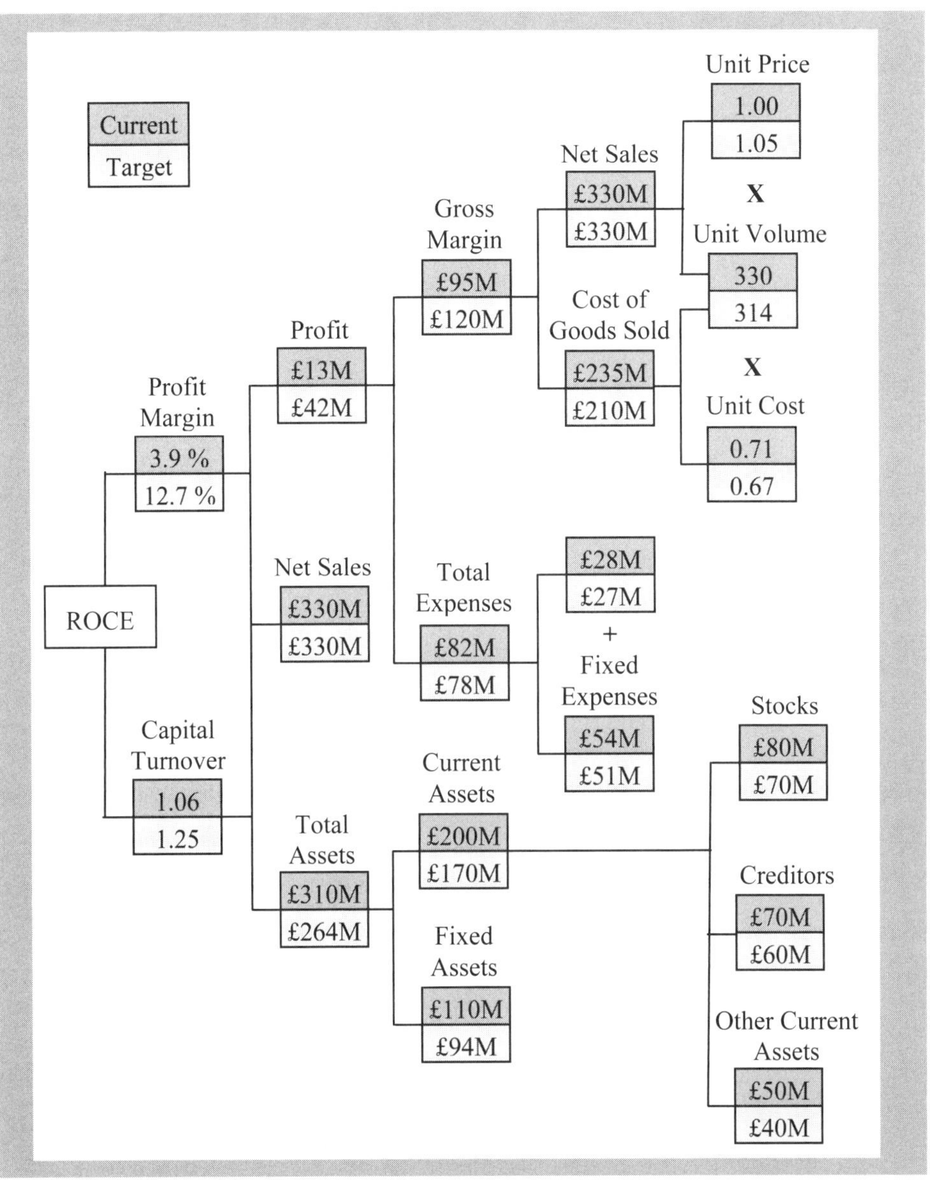

EXHIBIT 7.6 Chemco: Return on Sales and Asset Turnover Model

Source: Doyle, Peter *Marketing Management and Strategy* 2nd Edition, Prentice Hall Europe

Aligning the corporate strategy with these value drivers maximizes shareholder value creation. This model is from a financial perspective. The organization also needs to be in alignment with the financial strategy. In his book, "Contemporary Strategy

Analysis", Robert Grant aligned these financial value drivers with operational performance targets, Exhibit 7.7.

Establishing metrics derived from operational performance targets and performing routine management reviews of them ensures the organization is in alignment with the corporate strategy. It also allows the management team to understand their role in helping the organization achieve the corporate strategy. Time tracking of trends can identify which areas might need additional creative solutions or investment to achieve targets.

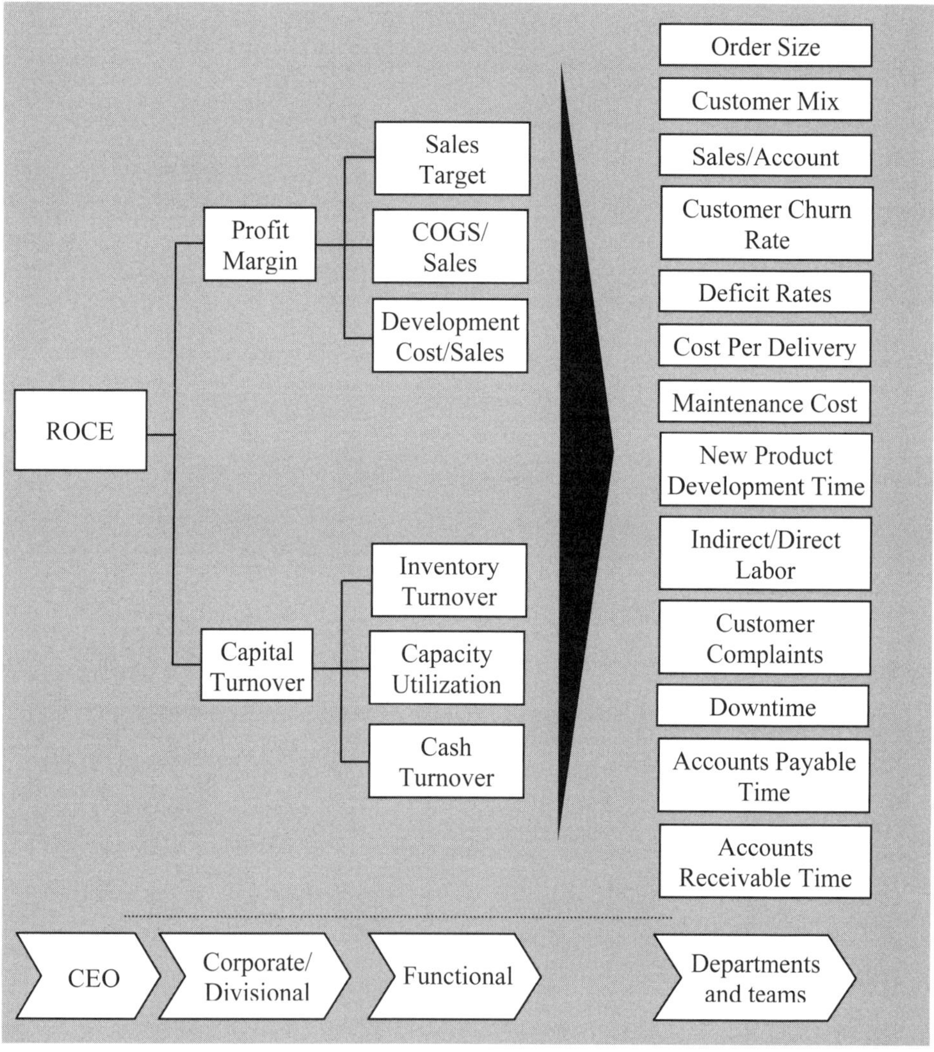

EXHIBIT 7.7 Linking Value Drivers to Performance Targets

Source: Grant, Robert M. *Contemporary Strategy Analysis* 3rd Edition, 1998, Blackwell Publishers Inc.

Comments from the Investment Community

Reviewing business plans and evaluating opportunities on a daily basis can be a very daunting task. Investors often claim there are not enough opportunities, while they are reviewing literally hundreds of business plans per year. Investments can fail to achieve funding due to aspects beyond technical. As part of an initial screen and throughout the due diligence process, the items of importance shared by investors are:

1. *People*: Top of the list – what is their history? Personal drive? Team fit? Integrity? Understanding of what needs to be done and ability to evolve with the company as it grows?
2. *Board*: Independence, ability to make tough decisions, mix of skills, ability to devote time to the company versus other commitments.
3. *Market*: Is it big enough to care about? Or could it become big enough? What is the market structure of incumbents; channels to end customers and who controls them today?
4. *Capital Efficiency*: How much cash will be needed? Can cash be feed in proportional to success [service-type play], or is a big bet up-front required [infrastructure-type play]?
5. *Exit Scenarios*: If things went well/badly, what would be the payback? Why would somebody buy the company? Who are the potential buyers and what is required to be demonstrated to catch their eye, in 3-5 years time?

A few investment organizations have expressed frustrations with common mistakes made by startup organizations and requested the opportunity to include a list of Do's and Don'ts when raising funds:

1. Do write a good, short business plan yourself. Don't pay a consultant to write a 200 page plan that no one will read and is full of the usual platitudes.
2. Do be very clear about the value proposition to the customer - why would someone want this good or service in this timeframe? Learn how to articulate the technology and value proposition and be prepared to recite it spontaneously.
3. Do be open and honest about the challenges and risks; no startup is risk free - far from it.
4. Do freely admit what you don't know - no one has all the answers. The phrase, "I don't know and will find out" will get more respect than making something up.
5. Do clearly spell out what you want and need - cash, resources, timing, valuation, VC types ... - nebulous requests are as bad as none and will waste time.
6. Do be reasonable about your demands - a little less cash or at a little lower valuation is better than none; remember, it's a marathon and if you demonstrate success in the first phase, you will be rewarded at the next financing round.
7. Do be ready to present the same slides 50 times and be asked the same dumb questions 100 times without looking bored.

8. If performing a spin-out: Do fully assign IP to the newly formed company. Don't license it with control checks and balances. Let the spin-out drive it to success. If the spin-out fails, let the IP ownership return to the parent company.
9. Don't formulate a business strategy that double-dips (charges twice) the customer. Companies tend to stay away from suppliers who charge them twice for a product or service.
10. Don't hide problems, tell them up front. Problems will be uncovered during due diligence and if not disclosed up front, investors will walk. Breaking of trust is the "Biggest deal killer".

Summary

Bridging the gap between investors and innovators is performed by improved communications. Understanding perspectives and expectations from both promotes a manageable and cooperative relationship, rallying behind unified objectives.

Investors approach an opportunity from their investment profile. This profile will be comprised of technology sector, limitation of capital, and risk management. Emerging technologies seeking funding will have a better chance of obtaining funding if the seek out investors for their sector. When a lead investor is identified, they will often communicate within the investment network to identify one or more co-investors to share the risk.

Throughout the investment cycles a manager of an emerging technology will be a continuously communicating with investors. While seeking funding from seed to IPO they will be negotiating valuations and term sheets, identifying and achieving milestones, delivering value to shareholders, and managing resources.

Cash flow management provided a high level approach to managing capital and provided insight on how to align the organization with the corporate strategy. As cash flow transitions to revenue generation, it is the CEO's responsibility to deliver the numbers. Shareholders will be pressing for a return on investment with efficient capital management. This can be achieved by aligning the value drivers for operations to the financial strategy. With the use of metrics derived from operational performance targets and the value drivers, the organization can be monitored.

The investment community has provided suggestions to help enhance the presentations for emerging technologies. They asked for clear definition of required cash, resources, timing, value proposition and valuation. They also provided a list of the top 5 critical items they consider before making an investment. There are plenty of publications and websites which publish similar content. Content designed to help startup CEOs become a successful business partner.

Chapter 8

Organizational Management

The more accountable I can make you, the easier it is for you to show you're a great performer... The more I use a matrix, the easier I make it to blame someone else. – Mark V. Hurd

Leadership verses Management

There have been numerous books and articles published on organizational management. These range from the traditional hierarchy structure to modern matrix structures, such as the McKinsey 7S model. In a startup organization it is the CEO who is held accountable for delivery of numbers and milestones. It is therefore imperative that the CEO carefully select the team and hold the final word on tough decisions; hierarchy structure. This does not imply that team building skills are not required. In modern organizations the successful CEO of a startup will prove to be more effective if they possess both leadership and management skills. A good manager does not commonly make a good leader; likewise a good leader commonly does not make a good manager. Each requires different skills with different objectives. Exhibit 8.1 identifies key differences between a leader and a manager.

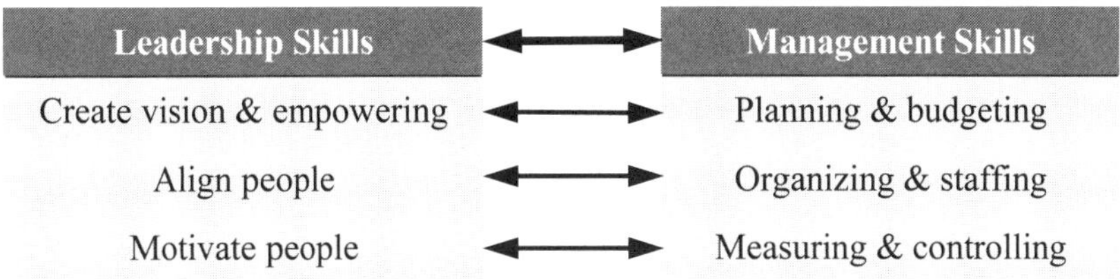

Exhibit 8.1 Skill differences between leadership and management

Leaders must use some management skills and managers must use some leadership skills. Understanding the differences and workings within their basic framework can make the workplace more productive. Focusing only on managing the day-to-day operations to ensure optimal performance, sacrifices efforts spent on future direction and strategy. This can result in zero or negative growth. Likewise, focusing only on strategy and direction without knowledge of organizational capability or performance can result in a strategy or direction that cannot be achieved. Again, this can result in zero or negative growth. In a startup role, the CEO will need to exhibit both leadership and management skills while building the organization. As the organization grows the management structure will need to expand and grow accordingly, empowering people at all levels throughout the organization.

Values

At the heart of every organization are its core values. It lays the foundation to the organization's culture. Core values reflect the corporation's intent and image to customers and to the local community. The values a firm embraces can help build employee commitment and loyalty, and offer the basis for differentiation when building relationships with other businesses. For example:

- ♦ **Ch2m Hill** was founded in 1946 by four individuals to perform engineering consulting services. Ch2m Hill was based on a strong foundation of doing the right thing, by their clients and their employees. Among the values held dearest are:

    - Honesty, integrity and trust
    - Responsive client focus
    - People-oriented
    - Quest for innovation and continuous improvement
    - Collaborative and enjoyable work environment
    - Sustainable, long-term growth
    - Community commitment
    - Challenging work opportunities and a strong work ethic
    - Commitment to safety, health, and environmental protection

    By 2007 Ch2m Hill's revenues had grown to more than US$5 Billion and achieved the industry ranking of number 1 in 10 different construction sectors. Ch2m Hill also made Fortune Magazine's "100 Best Companies to Work For" for the years 2003, 2006 and 2008.

- ♦ Founded in 1955, **McDonald's** ability to service 31,000 stores in 115 countries resulting in 2007 revenues of US$23 Billion cannot be explained by profit and strategy alone. McDonalds is sustained by a culture and principles of "quality, consistency, cleanliness and value."

Linking of strategy to the broader pursuit of social and moral purpose can facilitate rather than hinder long term performance.

Mission

An organization's sense of purpose can take the form of a vision that motivates its founding and sustains its development. A mission statement is comprised of two elements: The tasks or goals to be accomplished and the purpose or reason for executing those tasks. It encompasses what will be done and how to know when success has been achieved. Jack Welch set the divisional mission statement as "Be number 1 or number 2". This is a very brief and clear statement which defines the goals and provides a clear metric to determine if success has been achieved. This mission also includes clarity of value to stakeholders. It can also be referred to as a BHAG (Big Hairy Audacious Goal), defined by James Collins and Jerry Porras as, "A Powerful Mechanism to Stimulate Progress". BHAGs have been used throughout history to motivate and inspire people to achieve great success. John F. Kennedy used it on May 25, 1961 when he publicly announced, "that this Nation should commit itself to achieving the goal, before this decade is out, of landing a man on the moon and returning him safely to earth." This mission statement clearly identifies the task to be accomplished and the metric to measure if success has been achieved. It also places a timeline, creating urgency to the mission.

Referring to Collins and Porras' comparison example of mission statements between General Electric and Westinghouse, one is vague and the other is a clear message.

General Electric	Westinghouse
Become #1 or #2 in every market we serve and revolutionize this company to have the speed and agility of a small enterprise	Total Quality Market Leadership Technology Driven Global Focused Growth Diversified

Exhibit 8.2 Mission Statements of General Electric and Westinghouse

Source: Collins, James C and Porras, Jerry I (1997) *Built to Last*, HarperBusiness, A division of Harper-Collins Publishers

One can see that the Westinghouse mission statement does not clearly define the task to be accomplished; neither does it provide a metric to determine if success has been achieved. It does not create momentum or inspire people to drive toward a goal. Whereas the General Electric mission statement provides clarity to both goal and metric of success.

Some mission statements can on the surface appear to contain vague objectives. They can include statements such as, "Superior financial performance". This comment does not clearly identify the objective. It is measured by return on capital employed or earnings growth. But to have value it must be benchmarked against something; industry average, industry leaders, expectation of analysts? If an organization includes such statements into their mission statement, how it will be measured must also be made clear to the employees and shareholders. These metrics are to be benchmarked against leading industry organizations. Exhibit 8.3 identifies a few mission objectives which can appear to be vague in clarity yet identifies possible metrics which can be used to determine if success has been achieved.

Mission Objective	Metric(s)
Superior stockholder return	Total stockholder return
Superior financial performance	Return on capital employed Earnings growth
Most admired by customers	Customer satisfaction surveys
Competitive operational advantage	Operational expense per unit of product or service
Community admiration	Public favorability index
Committed team	Global employee and corporate survey results

Exhibit 8.3 Mission statement objective metrics

Source: Grant, Robert M (1998) *Contemporary Strategy Analysis 3rd Edition* Blackwell Publishers Ltd.

Defining a mission of *Competitive operational advantage* is a different message to employees and stockholders than a mission statement of *Lowest operational expense per unit of product.* Lowest operational expense is a tangible number that can be quantified. It is a value that an organization can track as they progress in their drive toward achievement. It provides clarity.

Resource & Team Management

The most critical aspect of managing an emerging technology is formulation of the team. This founding team must be comprised of individuals who bring a clear value to the organization. Establishing the team is an exercise of strategically searching the industry for qualified individuals who exhibit unique complimentary skills and values which are in alignment with the corporation. This is not an exercise of hiring friends and assigning them a new title. An early strategic marketing analysis will identify a target market segment. This revealed industry sector is good place to start searching for highly qualified resources to drive the innovation to market adoption.

The founding team is the group of innovators selected to transform the organization into alignment with the market, via the *customer discovery process*. For these members to successfully work closely together in a highly dynamic environment, it will require trust. Trust is an important element in team dynamics. When it is lost, a team's productivity and cooperation breaks down. They become dysfunctional. Carlson & Wilmot identified *The Trust Commandments*. These commandments are essential behaviors that build trust and forms social bonds that hold teams together.

The Trust Commandments

Respect for Others – a starting point for trust and productive collaboration. Respect is a demonstration that you value the contributions of others. If you do not respect others, they will not work with you and contribute to your success

Absolute Integrity – When respect and integrity are encouraged, a robust environment can be created where these values become the team's norm. Someone who does not follow through on agreements, is abusive, tries to trick you, pretends not to understand your legitimate business needs, or takes credit for someone else's contributions are clear indicators that this individual will be a poor partner.

Generosity of Spirit – Greed kills. We often ask, "Would you rather have 100 percent of noting or 1 percent of a million dollars?" For various reasons, there are numerous people who choose the small, suboptimal return, rather than collaborating with others to create a bigger opportunity.

"Trust is at the heart of teamwork" – Carlson & Wilmot

All three items bind a team together. They serve as the medium that allows the team to know they can rely on each other. Maintaining productivity and innovation is essential. Establishing guidelines, similar to these, as fundamental conditions of the working environment can create a solid foundation for success.

Culture is defined and demonstrated by the executive staff. This is exampled by Disney where employees are actors and customers are guests. Actors are carefully trained on how to communicate with each other, and to guests. This process helps create and maintain the corporate culture. Creating and nurturing a corporate culture is also exampled by one of Ch2m Hill's Founders and original CEO, Jim Howland, where in 1982 he wrote *Jim's little Yellow Book*. This book captured many of the values upon which the firm was built. Howland's thought 'N' expressed, "*Supervisors so often say, "I want you to do this" or "I want you to do that." Better to say, "It will work best if you do this" or "To conform to the firm's policies, please do that." We do things because it is good for the company, not because of an individual's desires.*" Communication coaching can serve as an effective tool to create and manage the organization's culture.

Professional High Performance Team

There exist organizations which specialize in training and developing standards for professionals. For example, many companies commonly promote a highly skilled technical person to the position of project manager, without proper training. These individuals are placed into a role different than the role in which they have a proven successful track record. They move from a mostly technical position to a position requiring technical and business level communications. Entering this new environment without training often ends with catastrophic results. Over the years this corporate behavior has resulted in very inconsistent project management skills with more failures than successes. In response to this, several international project management organizations have evolved, chartered to create training and professional certification programs, improving the quality and consistency of project management. An efficient and effective organization will leverage these professional certifications for enhancing the skills of their resources.

Professional certifications are designed to ensure that, through examination among piers, an individual is qualified to perform the associated function. Another example is the CBM (Certificate of Business Management). The CBM is a professional certification based on an MBA curriculum that is earned during, after, or in lieu of an MBA. It is a standardized exam based on the topics covered in the majority of MBA programs and is awarded to individuals who meet the eligibility requirements and pass the 16-hour, four-part exam. With the increasing number of universities producing students with an accredited MBA, there still remains inconsistency. The CBM Designation validates the mastery of business management knowledge, skills, and abilities and was developed by business practitioners to meet the needs of practicing business managers. Professional certifications help create global level standards and expectations within their field of expertise. It must also be clarified that a professional certification is not equivalent to an accredited degree program. An employee whose educational and training is exclusively based upon a certification of C++ programming will contain different skill sets than one who also holds a degree in computer science.

The list of examples where inconsistencies among professionals from varying industries have resulted in an evolution of professional certification programs is exhausting. Back in the early 1990's the IT (Internet technology) industry learned the value of employees holding network certifications. This reduced the inconsistent network configurations resulting networks going down and loss of productivity. Hiring employees with certifications, or training them, will help create a highly effective world class organization.

Managing FUD

As the customer development process unfolds, the organization can experience significant challenges. These challenges can result in required changes; changes in scope, direction, or application of the technology. Additional items such as adherence to regulatory requirements or international standards can require change. When faced with significant challenges or required changes, team members can begin to exhibit stress. They may be required to change the way they have done things for many years. They can be consumed with *Fear, Uncertainty* and *Doubt* (FUD), exhibiting resistance to change. A role of champions, and guidance to managers, is to identify FUD when it arises and transform it for success. FUD is commonly exhibited through anger or irrational behavior. The individual will utter complaints associated with the work environment. A competent champion can listen closely to team members and identify what is scaring them. Carlson and Wilmot provided the following list of FUD complaints and suggested *Reframe* that a champion will hear. The table provides a method of restating it in a way that demonstrates understanding and a sincere desire to address it.

Complaint	Reframe
"I'm too busy."	"I'm out of energy; give me some help"
"I can't work with her."	"She is different; how do I relate?"
"Doesn't this violate who we are?"	"How do I plug into this?"
"This won't work."	"I can't visualize it."
"I'll fight you to the death on this one"	"I feel excluded from the vision."
"Why do we have to work in teams?"	"Show me how to collaborate."
"No way will this be fairly done"	"Will I be rewarded?"
"Change, change, change – that's all we hear"	"I don't know how to adapt."

Exhibit 8.4 FUD complaints and suggested reframe

Potentially each FUD expression can be transformed into success. This is achieved by helping the team member articulate their needs, positively highlight what's blocking them, and working through the issues one by one. Perform this using verbal communication, not through email, motivational posters or position statements. Do this with the team members who share the issue. Generally the issues will not quickly disappear. But if addressed in a positive supportive manner, most will fade away.

Did NASA and the United States, as a country, experience Fear, Uncertainty and Doubt when John F. Kennedy set the mission to place a man on the moon within nine and a half years? Some level of FUD had to have been present and obviously managed. However, there is a difference between FUD and destructive behavior. Carlson and Wilmot also include the clarification of unacceptable behaviors. These unacceptable behaviors include cynicism, passive-aggressive resistance, criticism of team members to outsides, and end runs.

Cynicism – Cynicism is corrosive. Cynics question everyone's motives and the implied motive is always bad. Cynicism must be identified and met head on. If ignored, it can propagate throughout an organization bringing progress to a halt. Alfie Kohn argued that a person's cynicism stems from escaping responsibility. Cynicism can be countered by an enthusiastic, thoughtful, uncompromising champion. The team can rally around a change when it is clearly and consistently communicated with a shared vision. Communications and trust must be returned.

Passive-Aggressive Resistance – Some people believe that if they say nothing, they are free to walk out of the room and disagree with the conclusions of the group. This is obvious to those who see it. This type of individual will say nothing during a meeting and then sabotage the team by means of not returning phone calls or answering emails. These individuals typically sit in meetings demonstrating they are not engaged by folding their arms, leaning back on their chair, working on non-related tasks. They will vividly demonstrate they are not engaged. When this "I don't agree" body language is observed it must be faced immediately. Go around the room and ask for commitment to the decision.

Criticism Behind Someone's Back – This is one of the most toxic of all behaviors. It destroys team cohesion and lets others know you don't face and resolve issues. When performed to upper management it serves a purpose to discredit without the responsibility of reciprocation. Common phrases spoken behind someone's back are, "Vickie just doesn't seem to be a team player." and "Ralph is slowing down the team. He doesn't know what he is doing." These can be devised as an attempt to make one appear to be "a team player" and "a pier to Ralph". Criticism of others deflates the strength and purpose of a team. Behind the back criticisms should never go unchecked.

End Runs – In grade school tattletales made the worst kinds of friends. As a result of a disagreement, Steve would call Johnny a bad name. Johnny would respond in kind. Then Steve would run to the teacher saying, "Teacher, Johnny called me a bad name." An end run is when you take your criticism around a person and directly to their supervisor. Those who perform end runs often justify them claiming it was the only way to obtain change. This is usually self-justification for poor behavior, fear of conflict, and inadequate interpersonal skills. End runs result in loss of trust.

Unfortunately these behaviors run rampant within the culture of many organizations. Culture is a direct reflection of the leadership. It is the responsibility of an organization's executive team to train managers to identify these unacceptable behaviors and address them immediately.

Robert Slater published Jack Welch's characterization of employee types. Welch placed employees in four categories:

A. *Delvers on commitments – financial or otherwise – and shares GE's values.* "His or her future is an easy call," says Welch. "Onward and upward"
B. *Does not meet commitments and does not share GE's values.* "Not as pleasant a call, but equally easy."
C. *Misses commitments but shares the values.* "He or she usually gets a second chance, preferably in a different environment."
D. *Delivers on commitments but does not subscribe to GE's values.* What happens to managers who deliver the numbers but do not live the GE values? According to Welch, they get fired.

Welch's rules:

- Eliminate employees who do not live the company values, even if their numbers are good. Difficult, yes, but absolutely necessary.
- Nurture the employees who live up to company values, even if they don't make their numbers. Consider reassigning them if their numbers continue to falter
- Give employees more responsibility, and they will make better decisions. By making your employees more accountable, you make your organization more productive.

What does this add up too? It means surround yourself with A's, *the best people.* At times, human resource management can be difficult. Remembering to place the organization first will result in the creation of a productive organization.

Organizational Expansion

As the emerging technology migrates through the different stages of development, the organization has to grow accordingly. It is the business strategy that will determine the growth rate and potential size of the organization. If the plan is to build and directly compete in the market in a global scale, then the comprehensive operational development and organic growth to support the plan must occur. Outsourcing and collaborating will reduce direct head count and burn rate. These should be given serious consideration to avoid "Empire building". Occasionally, after receipt of funding, startup companies can build an organization in preparation for the pending work load. They could prematurely create a large head count, resulting in an unsustainable high burn rate. This practice is referred to as "Empire building". A significant amount of the early research and technical development tasks can be outsourced, reducing the time to achieve milestones. Outsourcing provides immediate access to resources which otherwise could take significant time to hire and provide development tools.

The CEO must plan several steps ahead of the organization. As each technical milestone is achieved, how will the subsequent stage of technology development occur? Which functions will be developed organically? Which will be outsourced? Growth planning correlates to funding and revenue. What will be required to reduce the time to market, generate revenue, and maintain strategic control?

As the organization grows, another significant item to consider is the rule of 150. As described by Malcolm Gladwell, "The rule of 150 suggests that the size of a group is another one of those subtle contextual factors that can make a big difference. Once that line, that Tipping Point, is crossed, humans in a functioning social group begin to behave very differently. The rule of 150 says that congregants of a rapidly expanding church, or the members of a social club, or anyone in a group activity banking on the epidemic spread of shared ideas needs to be particularly cognizant of the perils of bigness. Crossing the 150 line is a small change that can make a big difference. Below 150, orders can be implemented and unruly behavior controlled on the basis or personal loyalties and direct man-to-man communication. With larger groups this becomes impossible. Through centuries of development and refinement, he rule of 150 has been adopted by organizations such as the Hutterite colonies, the Methodist church, and Gore Associates. The military uses 200 as the limit for functional fighting unit. With groups over 150, complicated hierarchies, rule, regulations and formal measures are imposed to command loyalty and cohesion. Below 150, it is possible to achieve the same goals informally.

Social groups which reach the number of 150, on their own, begin to form clans or sub groups. This can occur within an organization, especially when the strategy includes more than one business unit or market sector. When the organization head count achieves 150, the long term business strategy might be better served by creating two divisions. Many acquisitions incorporate operational cost reductions as part of the justification. There are common claims that resources like purchasing and accounting can be combined to reduce operational costs. This also implies that the purchasing and accounting groups of both organizations were only tasked at 50% capacity. When the rule of 150 is factored in, and ignored, it can serve as a natural social ingredient for failure.

Summary

Avoid the cross functional matrix management. As described by Mark Hurd and Jack Welch, authority and the ability to execute come with accountability.

Execute the Plan, build and monitor the business, drive to adoption. Monitor market, technology trends and emerging threats. Disseminate information to stakeholders and employees.

Team and stakeholder management is an ongoing task. Hire the best people. Occasionally friends are hired for key positions with good intentions. This can place them in a role for which they may contain little or no experience. If they don't have the skills, train them, utilize executive training programs from leading universities. Leverage professional ceritifications to identify high performance employees. Don't start "Empire Building". Keep the resources realistic and balanced.

Remind employees of the value proposition your organization provides to the customer and their role in helping the organization achieve its goals.

Train and monitor executives and managers with respect to corporate culture and communications. Be cautious of distractors to reaserch and development, and to the corporate goals. Utilize tools outlined in the first couple chapters to develop metrics and monitor the business. Ensure R&D is both Value Proposition to the customer, and IP generation focused.

Chapter 9

Corporate Perspectives

"We are all victims of our own socialization. The lenses through which we perceive the world are colored by our own ideology, experiences, and established management practices." – C. K. Prahalad

Institutional Memory

"The Phonograph has no commercial value at all" – Thomas Edison

"Such startling announcements as these should be deprecated as being unworthy of science and mischievous to its true progress" – Sr. William Siemens, on Edison's light bulb

Institutional memory occurs when the same people, attend the same conferences, with the same technologies, to network with the same people, working for the same company, for several years. Over time they become highly valued experts in their industry. Their knowledge of relationships between corporations and products within the industry carries a significant amount of value. However, the problem which commonly arises as a result of this process is a loss in the ability to recognize emerging opportunities, and more importantly the failure to recognize the early warning signs of value outflow within their own organization. A major contributor to why corporations fail to respond to value outflow is, "Institutional Memory". Management typically responds with justifications why customers and key personnel are leaving. Citing everything from a poor economy, to a comment recently heard from the CEO of a US$15B corporation, "It is okay for our top researchers to leave. It helps improve competition in the industry".

Does the organization have the ability to recognize an emerging opportunity? Is management focused only on obtaining the "Big Win", a business model projecting revenues above $1 Billion? Ignoring the *Crossing the Chasm* effect of business evolution or technology adoption? Are your technology leaders' nay-saying emerging technologies, "We tried that 20 years ago, it doesn't work"?

There is an old Korean parable, "A turtle cannot move if he doesn't stick his neck out". Corporations throughout the globe are comprised of managers who are either risk takers or non-risk takers. The non-risk taking manager thrives on the theory that no risk equals no mistake. No mistakes make it easier to be promoted as corporations frequently do. Over time this can fill the top management with overly risk adverse individuals poised to take no risk and subsequently no growth. A common phrase utilized by higher level risk adverse managers is, "Look at how much money I saved by shutting down all these investments". Alternatively, the risk taking manager envisions potential opportunity and drives toward it. If the risk result is negative to the corporation, they often find themselves suddenly in search of new employment. Corporations around the globe frequently promote the non risk taker and punish a failed risk taker who obtains valuable lessons from each attempt. The result is a corporation filled with decision makers who are non risk takers, holding on until retirement.

It is important for executives to identify and respond to changes in the industry. Corporations who want to maintain, or obtain a competitive advantage have committed their executives to continued education. The single most important decision a corporation can make in avoiding "institutional memory" is to make the commitment of continued education. Giving their executives the latest tools designed to help them make better-informed decisions, more importantly it avoids complacency and improves the quality of Information Intelligence.

Product Portfolio Management

Knowledge of a firm's and/or competitor's product portfolio and brand portfolio strategies plays a role in evaluating emerging technologies. Identifying these trends prove to be highly valuable when identifying collaborators or demonstrating value to stakeholders. Just because the emerging technology appears to have a successful future, does not mean the opportunity should be pursued. Successful organizations have a defined strategy and direction. Opportunities derived from emerging technologies need to be in alignment with corporate strategies. Additionally, chasing too many opportunities can result in diluted resources. Attempting to pursue too many opportunities results in projects being under staffed and under supported by upper management. The organization could lose focus and direction resulting in missed opportunities and possible dilution of brand. Most publicly traded companies succumb to the short-term pressures of the stock market to spend equity reserves and increase debt. When not managed properly these practices can remove available funds required for investing in emerging technologies, and can run an organization so lean there are no resources available to pursue a critical opportunity, or sometimes survive an economic hardship. Companies, which are highly leveraged commonly, reduce budgets for research and development and other costs for which the return of value is in the future. An organization needs to have resources available to respond to rapid market changes or market intrusions. These pressures require executives to work not only more efficient and smarter, but also in tighter collaborations within the organization. This results in a unified understanding of the organization's product strategies, brand strategies, their relationships, and maintains direction and focus.

In "*Portfolio Management for New Products*" Cooper *et al.* formerly defines portfolio management as the following:

> "Portfolio management for new products is a dynamic decision process wherein the list of active new products and R&D projects is constantly revised. In this process, new projects are evaluated, selected, and prioritized. Existing projects may be accelerated, killed, or deprioritized and resources are allocated and reallocated to the active projects. The portfolio decision process is characterized by uncertain and changing information, dynamic opportunities, multiple goals and strategic considerations, interdependence among projects, and multiple decision makers and locations." (P.3)

Portfolio management assists the organization in prioritizing and aligning resources to a defined strategy. Organizations, which make decisions without using a portfolio management method often experience "cause and effect" behavior as described by Cooper *et al.*, Exhibit 9.1.

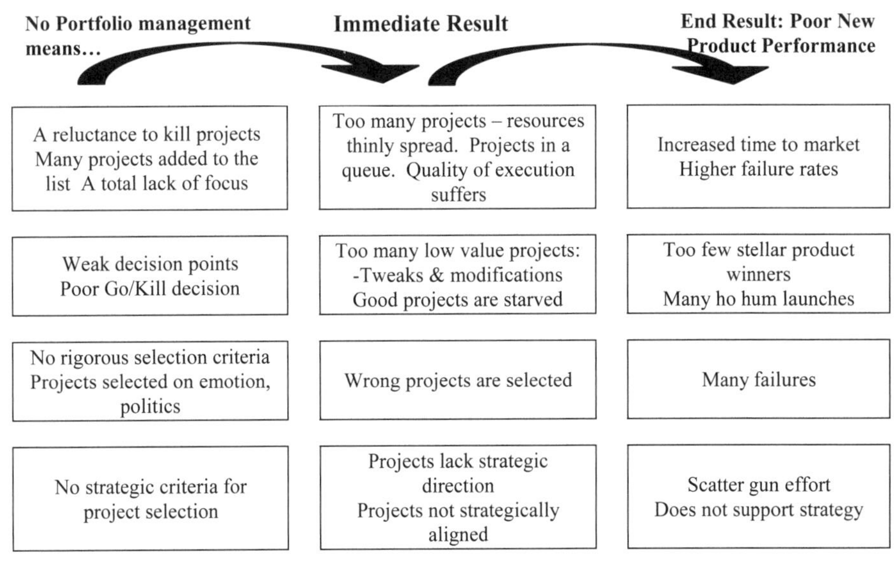

EXHIBIT 9.1 What happens When You Have No Portfolio Management Method

Source: Cooper et al, "Portfolio Management for New Products" (P.5)

Mapping market trends and emerging technologies allows an organization to allocate core competencies and resources required to position themselves in emerging markets. Using tools such as bubble diagrams similar to Cooper *et al*, or radar diagrams as Slywotzky an organization can use this information to make effective strategic decisions.

Cooper *et al.* developed a Gate process designed to move new products from concept to launch, which is inclusive of generic composite score card models for rating and ranking new product projects, Exhibit 9.2. Scorecard models are used when moving a product from one phase to another. Based upon extensive research into critical success factors, Cooper *et al.* determined the following six major factors were critical for success (p.55):

1. *Strategic alignment and importance:* Is the project aligned with strategy, and is it strategically important? [Corporate values?]
2. *Product and competitive advantage:* Does the product offer unique customer benefits? Meet customer needs better than competitors? Provide good value for money?
3. *Market attractiveness:* Is the target market an attractive one – size, growth, margins, and competition?
4. *Leverage core competencies:* Does the project build on strengths, experiences, and competencies in marketing, technology, and operations?
5. *Technical feasibility:* What is the likelihood of technical feasibility – size of gap? Complexity? Uncertainty?
6. *Financial reward:* Can this project make money? How sure are we? Is it worth the risk[s]?

The addition of corporate values to Strategic Alignment and Importance reflects the organization's responsibility to sociological and environmental issues. Seiko Epson Corporation in Japan consistently express their core values of "Co-Existence" with the environment and "Saving Energy". These values are included in the evaluation process of undertaking business opportunities. Numerous times an opportunity is killed simply because it does not support the corporate core values. In words of Jack Welch, "The hardest thing in the world is to move against somebody who is delivering the goods but acting 180 degrees from [your values]. But if you don't act, you're not walking the talk and you're just an air bag." According Welch, managers who deliver the numbers but do not live the GE values get fired.

The rating system allows an organization to set acceptable limits for projects to be pursued. Note that care must be taken when filling in the rating column, to prevent personal bias from creeping into the metric.

Rating Scale					
Key Items	0	4	7	10	**Rating**
1. Strategic Alignment & Importance ✓ Strategic fit and importance ✓ Fits our strategy ✓ Important to do ✓ High impact on our business	Product not in alignment with or important to our business strategy; low impact: KILL	Somewhat supports business strategy; not too important; modest impact	Supports business strategy; important; good impact	Product aligns well with our business strategy; product very important to strategy; high impact	
2. Product and Competitive Advantage ✓ Unique customer benefits ✓ Value for money ✓ Customer feedback in next phase	None; negative or neutral customer feedback; poor value	Limited; marginally superior; fairly neutral feedback; OK value	Some new benefits; somewhat superior; good value; positive feedback	Major new benefits; very positive customer feedback; great value	
3. Market attractiveness ✓ Market size and growth ✓ Margins ✓ Competitive situation	Small or non-existent market; low growth & low margins; tough competition: KILL	Modest market; limited growth; fair margins; competitive	Significant market; good growth; good margins; modest competition	Large, growing, attractive market; good margins; weaker competition	
4. Leverages Core Competencies ✓ Technology ✓ Production ✓ Marketing & distribution/sales	No opportunities to leverage competencies; required skills/ experience/ resources strengths are weak: KILL	Some opportunities to leverage our competencies; our skills/ experience/ resources are modest	Considerable leverage possible; skills/experience needed for projects are within company	Excellent leverage of our strengths & competencies; excellent fit between project needs; our skills, experience, resources	
5. Technical Feasibility ✓ Small technical gap ✓ Not too complex technically ✓ Uses our in-house technology ✓ Demonstrated technical feasibility	Low; big gap; new science; technology new to company; have not been able to demonstrate technical feasibility: KILL	Modest; fairly large gap; quite a few hurtles but do-able; technology fairly new to company; limited evidence to support technical feasibility	Good; small gap; some hurtles, but attainable; have some evidence of technical feasibility	Straight-forward; largely engineering repackage; we have technology in house; have demonstrated technical feasibility	
6. Financial Reward vs. Risk ✓ Sizable, excellent opportunity ✓ Payback, NPV & IRR Ok ✓ Certainty of estimates ✓ Not too risky and difficult to do	Poor, limited opportunity; NPV negative, payback > 5 yrs; difficult to make money here; risky & tough to do: KILL	Modest opportunity; NPV positive; Payback = 4yrs; fairly difficult to make money; fairly risky & tough to do	Fairly good opportunity; NPV positive & good; payback = 2 yrs; probably can make money; modest risk and difficulty	Excellent opportunity; NPV positive & high; payback < 1 yr; not too risky & difficult to do	

EXHIBIT 9.2 Composite Best-Practices Criteria Scorecard

Source: Cooper et al, "Portfolio Management for New Products" (P.54)

Brand Portfolio Management

No selection criterion is complete without the inclusion of Branding. Brand portfolio management follows a similar process to product portfolio management. Decisions on branding are critical as it affects the overall value of the organization.

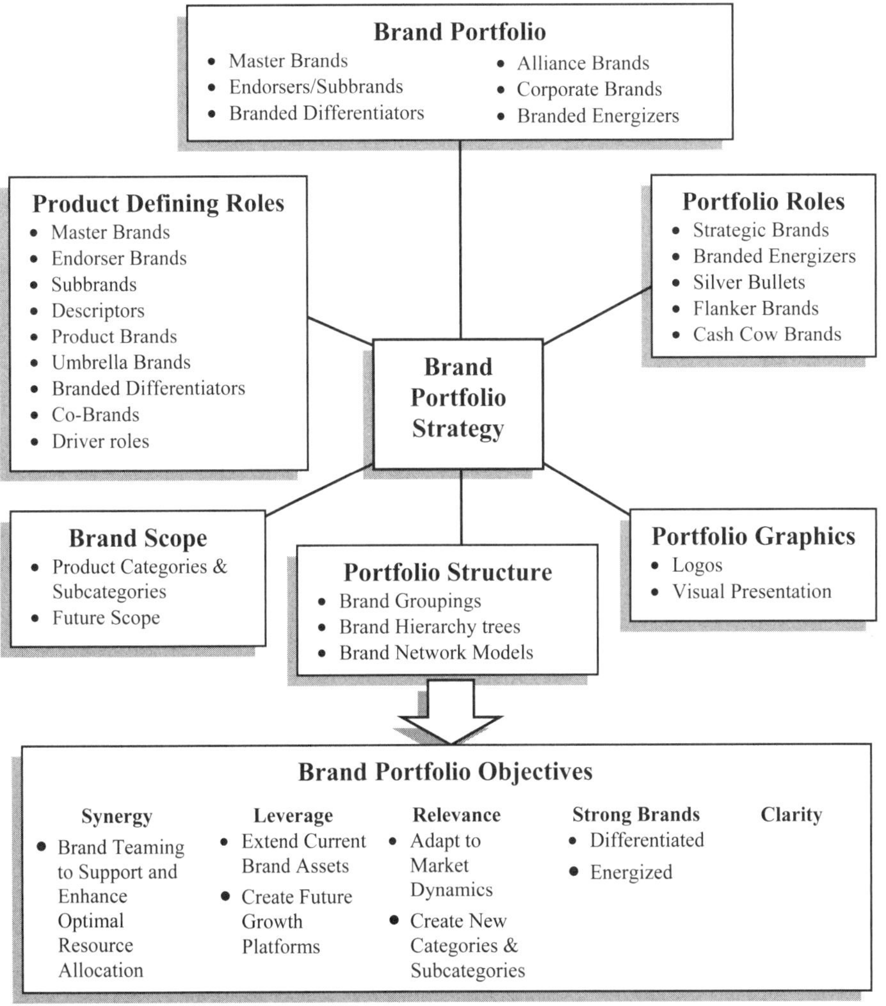

EXHIBIT 9.3 Brand Portfolio Strategy

Source: Aaker, David "Brand Portfolio Strategy" (P.17) (Reprinted with permission)

Appropriate product positioning and branding are essential in the decision/selection process. Rouge product launches have the ability to dilute brand value. A branding portfolio strategy will include products designed to provide specific branding functions. For example, functions of a sub-brand can be to introduce new offerings, stretch a master brand to compete in new markets, and serve as descriptor and driver roles. Including brand portfolio strategy into the selection criteria can enhance corporate portfolio performance. Aaker (2004) identifies six dimensions of a brand portfolio and represents them graphically in Exhibit 9.3.

Measuring Technology Performance

The conventional technology S-curve described by Christensen identifies the need for corporations to monitor the relationships between R&D efforts and the performance yielded as a result of those efforts. Failure to manage investments, forecasting, technology mapping, R&D, and corporate strategy planning precipitates in failed attempts. The technology S-curve depicted in Exhibit 9.4 deposes a theory that the rate of performance will be relatively slow in the early stages. The rate of performance increases as the technology gradually becomes adopted until it reaches a mature stage at which the performance becomes asymptotic.

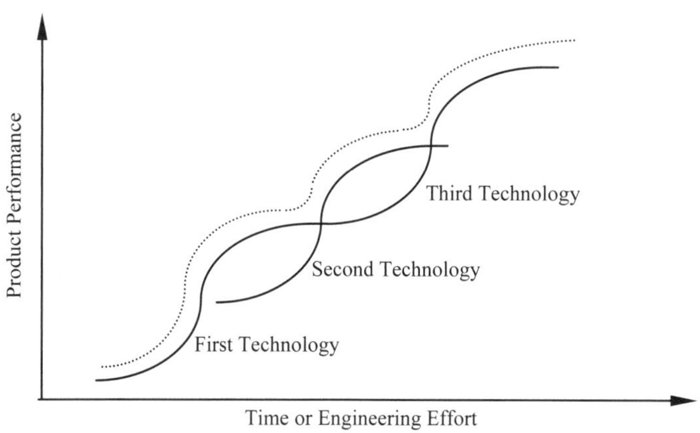

EXHIBIT 9.4 The Conventional Technology S-Curve

Source: Clayton M. Christensen, "The Innovator's Dilemma" (P.40)

At the point of inflection, a leading organization will have a next level technology positioned to emerge resulting in an increase of the performance. Effectively executed, over time this process increases the overall performance of the organization and is depicted by the dotted curve paralleling the introductions of subsequent technologies.

Adrian Slywotzky breaks down the business design profit curve into three phases of Value Migration. As seen in Exhibit 9.5, a business or technology progression undergoes three fundamental phases: First is the value inflow stage, followed by stability and finally value outflow.

During the Inflow phase, the organization is investing in R&D, mergers, joint ventures, or any other business strategy with the expectation the result will yield an increase in value to the organization. Benefits of first mover position, limited competition, and high growth rates accompanied with high profits are enjoyed.

During the stability phase, a limited number of competitors enjoy a stable market with reliable returns. Then during the outflow phase, more competitors arrive, driving prices down resulting in loss of market share, revenues and profits. This is the phase when technologies become commodities.

The inflow phase is heavily burdened with investment. Profits during the stability phase are obtained by improving processes and taking advantage of economies of scale. Eventually competition drives the profit out and corporations find themselves shipping product at or below cost, the outflow stage. Shipping at or below cost is frequently justified in an attempt to maintain market share, increase sales of other products, or to reduce inventories.

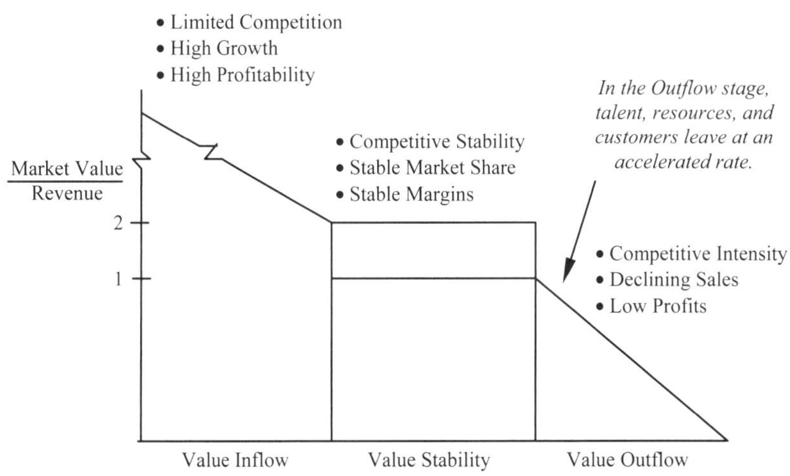

EXHIBIT 9.5 Three Stages of Value Migration

Source: Adrian Slywotzky, "Value Migration" (P.50)
(Note: Market Value = (Shares Outstanding X Stock Price) + Long Term Debt)

The inflow stage for memory chips was highly profitable. As the technology migrated to the stability phase, more manufacturers came online and enjoyed a profitable market until the memory chip became a commodity. Too many manufacturers flooded the market driving the prices down. Improving efficiencies in manufacturing and increased volumes were the only way to make a profit. Today, the memory market is a commodity market. In today's market it would not be a wise decision for an organization to invest in a semi-conductor factory for the purpose of manufacturing memory chips utilizing current technology. Even thought such an investment could result in a significant increase in revenue to the organization. There would be very little or no profit, no value, unless such a decision was founded upon an emerging technology such as FeRAM as an example.

There are early indicators of technologies and business designs entering the value outflow stage. Talent, resources and customers leave at an accelerated rate for example. The number one reason why corporations fail to respond to value outflow is, "Institutional Memory".

This three-phase life cycle for business designs can also be applied to technology migrations. Profits generated (Value Migration) and performances measured (Technology S-curve) are related. Exhibit 9.6 graphs the profits generated by a business design across the three stages of Value Migration. This graph represents the investments and returns on investment as the business design progresses through its life cycle.

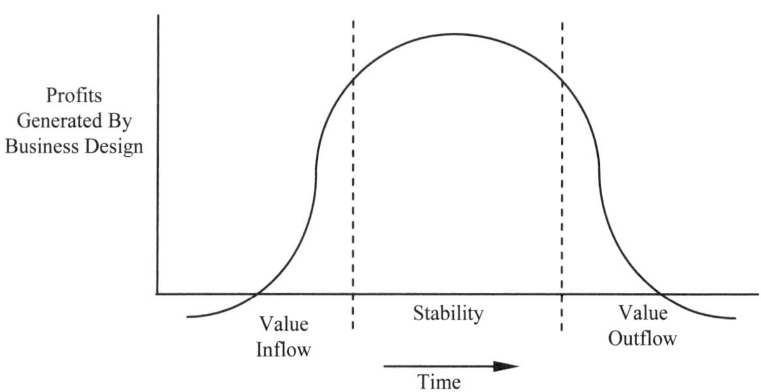

EXHIBIT 9.6 Business Design Profit Curve

Source: Adrian Slywotzky, "Value Migration" (P.49)

By retracing the profit curve at the point of inflection identifies the similarities and relationships between the business design profit curve and the technology S-curve as shown in Exhibit 9.7.

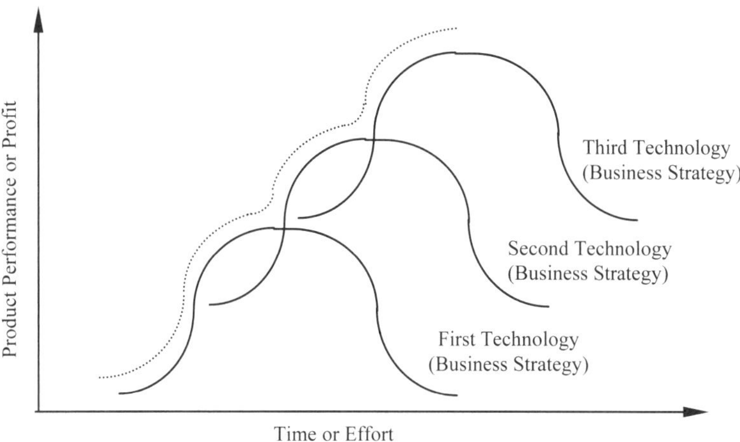

EXHIBIT 9.7 Relationship between Value Migration and Technology S-curve

Develop Organically or Acquire

Within many organizations there exists an opposing dichotomy. The struggle resides between two fundamental requirements for success. One is a requirement to incorporate leading technologies into products to obtain differential advantage in the market place, and the other is a requirement to protect product manufacturing by not allowing any single source suppliers. How does one incorporate leading technologies, which are patented, while simultaneously not allowing single source suppliers?

Many companies over the years have struggled internally with this problem. Solutions frequently adopted are: 1) organically develop a solution, 2) acquire the technology, or 3) Investment in the supply chain. Other solutions implemented have been investments in the supply chain and/or relaxing the single source requirement with the implementation of a risk management process.

Not solving this dichotomy, only results in products with old technologies.

Summary

Established organizations are planning for the future. They are managing their brand and associated offerings to support those brands. Techniques used to maintain a competitive edge can include creating a sub-brand which equals a competitor's luxury brand. Every opportunity is evaluated for value. Value can be individually measured with any number of metrics such as ROI, market share or environmental impact.

As an emerging technology is proposed to an incumbent market leader, one must understand the evaluation perspectives the organization is undertaking. The organization has an existing long term product portfolio strategies. They do not have engineers and scientists sitting around waiting for your technology to come along. Market leader organizations receive an extensive number of collaboration requests. Those which align with the existing strategy commonly receive attention.

Technologies associated with an opportunity are performance measured against the market and competing technologies. Significant reductions in costs are required. Marginal reductions can be achieved through production enhancements and market pressures.

Business strategies and market dynamics are constantly changing, another tool leading companies are using to maintain their competitive advantage is continued education. They are taking advantage of the emergence of executive education programs. It is in these programs where new business methodologies are taught in short time frames, reducing time from work and refreshing the knowledge base.

Chapter 10

Evaluation of Emerging Technologies Process

Many ideas grow better when transplanted into another mind than in the one where they sprang up" – Oliver Wendell Holmes

As mentioned in the previous chapters, an emerging technology begins with eureka. Thereafter, it is when the hard work begins. Exhibit 10.1 outlines a process for effectively evaluating an emerging technology while providing stakeholders the necessary information to make informed decisions.

The process comprises of five key phases. Each phase highlights critical issues, risks and milestones to be addressed. It identifies when to ask the right questions, when to perform technology and market analysis, and when to develop and execute the business strategy. It provides a method of analyzing where in the development cycle the emerging technology has progressed in terms of growth from concept to market adoption.

While Exhibit 10.1 demonstrates five independent phases, one should be aware that in practice, this rarely occurs. The evolution of an emerging technology more often requires cross-phase developments. Throughout the development of an emerging technology, markets, technologies and collaborations change. Breaking up the required tasks into a stepped phase process simply enables the technology to be evaluated and managed from a structured approach. Some phases have repeated items such NDAs, risk and business design. As the technology matures and the market dynamics change each must be revisited for review. This chapter will focus on the different aspects of the process.

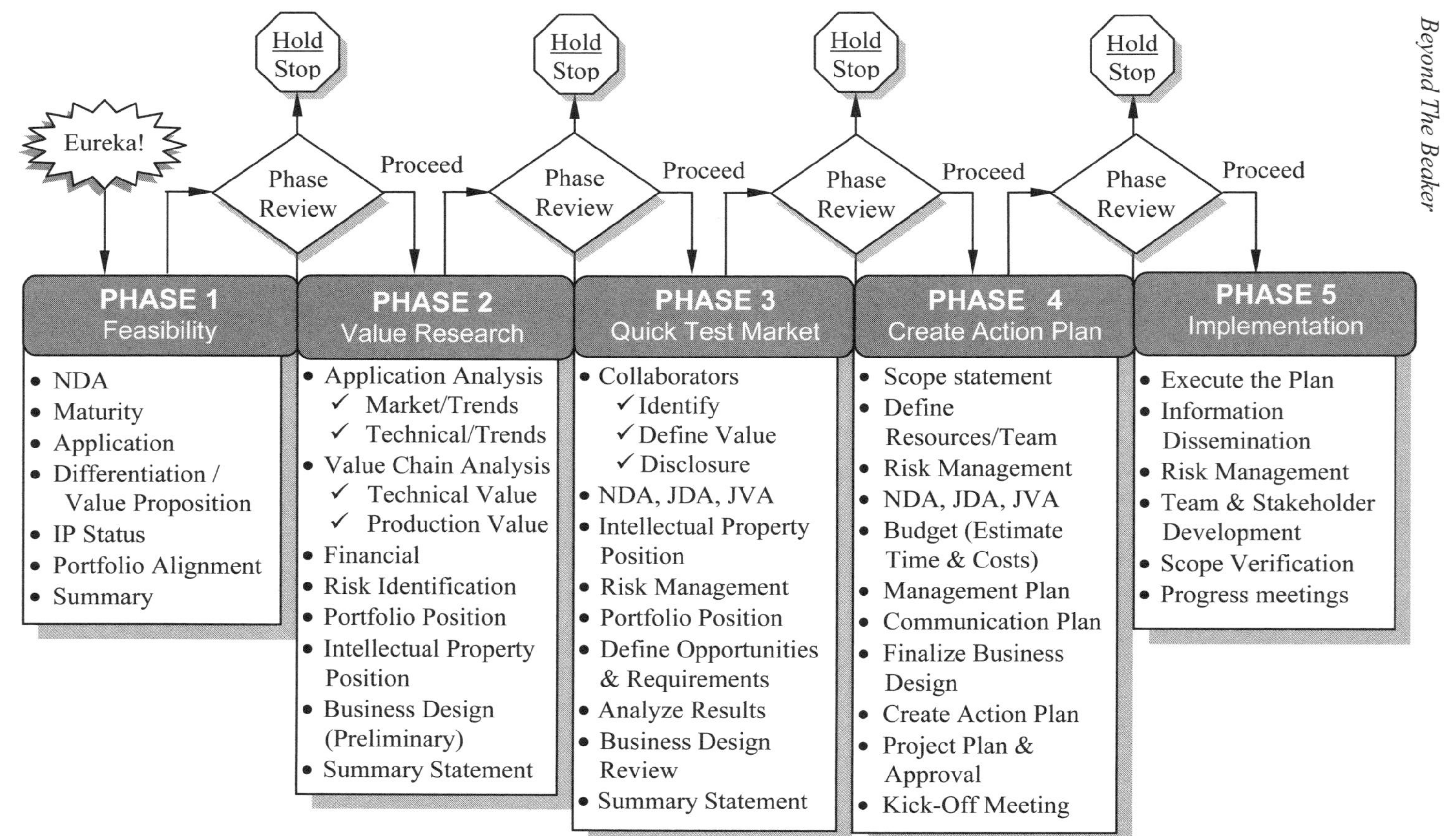

EXHIBIT 10.1 Evaluation of emerging technologies process

PHASE 1 - Feasibility

PHASE 1
Feasibility

- Maturity
- Application
- Differentiation/ Value Proposition
- IP Status
- Portfolio Alignment
- Summary

Phase 1 is the feasibility phase. This is the stage that identifies what the technology is comprised of, removes the hype around it, determines its value proposition and differentiation, its level of maturity, and the potential applications.

Significant time may pass between the conceptualization of an idea until the time it is evaluated. Late evaluations of an emerging technology can be accompanied by hype and/or speculation. Completing a comprehensive feasibility analysis can look past the hype and speculation to allow stakeholders the ability to make informed decisions.

This phase uses a scorecard model to benchmark the emerging technology against other technologies competing for resources. Each parameter listed under Feasibility will be discussed along with the associated rating system.

Maturity

How mature is the technology? Has it been demonstrated or is it only conceptual? A fundamental understanding of the technology is required and helps provide insight into the probability of achieving success. The rating system below places the maturity of the technology across the spectrum from a concept to a developed production technology:

Rating	1	2	3	4	5
Maturity	Conceptual	Laboratory testing	Obtained collaboration	Functional prototype	manufacturability issues resolved

- *Rating 1: Conceptual* Ideas at the conceptual level carry a higher risk of failure.
- *Rating 2: Laboratory Testing* Rigorous laboratory tests will minimize most major technical risks increasing stakeholder confidence.
- *Rating 3: Obtained Collaboration* When the fundamental principals are demonstrated through testing, the technology will begin to attract positive interest from potential collaborators - investors to technology partners.
- *Rating 4: Functional Prototype* Creating a functional prototype alleviates most high-level technical risks and demonstrates the technology has a viable application. The remaining work is typically process development in preparation for a market driven application.
- Rating 5: Manufacturability issues resolved

Application

Does the technology have an application? Often an emerging technology has no application or can be applied to a limited number of applications. Nanotechnology is an example. During the early stages of nanotechnology developments, companies demonstrated numerous capabilities and publish them with the anticipation that someone will respond with an application. Processing and functionality were advanced beyond the industry's ability to utilize the technologies. The rating of application correlates to market size of the application and a company's strategy.

Rating	1	2	3	4	5
Application	None	A "Me too" Technology, no significant differentiation, limited cost reduction	Moderate market, compliments or supplements existing technologies	Large market, competes against existing technology	Strategic correlation, large market

A difficult aspect to this analysis is the awareness of other emerging technologies positioned to compete within the same market. Occasionally technologies are developed to address a technical need which is inferior to other emerging and more mature technologies. When such an event occurs, it places a "Me too" rating on the emerging technology.

Differentiation/Value Proposition

Technology differentiation, what makes this technology different? What is its value proposition? Evaluating the attributes requires understanding how it is used and what role it plays in the technology value chain. Benchmarking against established technologies adopted by the market, their limitations, and projected directions of growth enables the ability to provide accurate ratings.

Rating	1	2	3	4	5
Differentiation/ Value Proposition	None	"Me too" Technology with limited differentiation	Superior performance at a lower cost	Replaces existing technologies at a lower cost	Disruptive Technology

IP Status

Intellectual property is fundamental to maintaining market share and executing business strategies. Tough questions need to be asked such as: Has the intellectual property for the technology been received? Applied for? How many patents filed and for how many applications? Does the intellectual property encompass enabling technologies and processes? Does it rely on complimentary intellectual property owned by another organization? If a company or organization controls the IP, does IP opportunity exist for the zones to market as described in Exhibit 6.1? What is the controlling company's business model or licensing model?

Rating	1	2	3	4	5
IP Status	None	Relies on other core patent(s)	Core obtained, Application and Processing filed	Core obtained, Application or Processing obtained	Core, Application and Processing obtained

Portfolio Alignment

How well does this technology adhere to the current corporate product/technology strategy or road map? Does it support the corporate branding strategy? Emerging technologies will either align with existing strategies, enable new opportunities, or be passed over. A decision to pursue a non-aligned technology can require significant additional resources and investment.

Rating	1	2	3	4	5
Portfolio Alignment	None	Redundant effort, upgrade or "Low hanging fruit"	Moderate fit with Branding and Product Strategies	Partial fit, compliments strategy and portfolio, support utilization	Fits within the existing Branding and Product Strategies

Phase 1 Summary

A brief summary statement provided by the analyzing individual or team, which includes a professional opinion of the technology, needs to be provided. This section also provides additional explanations of the ratings as needed. As projects and opportunities are reviewed over time, this summary statement will help provide insight into the next phase's evaluation. Summing the total score serves as a benchmark for evaluating a "Go", or "Stop/Hold" decision, or ranking and rating system. Every organization will develop an acceptable rating level for projects to pursue. Holding a phase review completes each phase. A team of decision makers is assembled and they perform an evaluation of the technology and associated phase contents. If the technology is authorized to proceed to the next phase, milestones the technology must overcome are to be identified. The identification of milestones for the technology is independent of the

Go, or Stop/Hold decision process. These milestones can be technological or business related: Such as meeting a lifetime specification, or having a collaborator on board. Documenting a reason for the Stop/Hold decision will identify what has to occur for the technology to become viable again. As time progresses, technologies can mature or morph into an alternative application, as demonstrated by the Gartner Hype Cycle during the Trough of Disillusionment.

The summary statement is where business model considerations or concerns need to be addressed. Sometimes cultural differences can exist, relationships, or collaborations preventing the adoption of an emerging technology resulting in a stop/hold decision at the phase review.

The table in Exhibit 10.2 represents an executive summary for the Phase 1 Feasibility assessment. It provides a column, which identifies feasibility parameters, comments about the parameters, and a rating column.

Phase 1 – Feasibility Table		
Parameter	**Comments**	**Rating**
Description	Describe Fundamental technology	- - -
Differentiation/ Value Proposition	What differentiates the technology	1-5
Maturity/Feasibility	Describe	1-5
Application(s)	Describe sectors of the technology	1-5
IP Status	Filed and how many? Requires other license?	1-5
Portfolio Alignment	Align with corporate core competence?	1-5
Summary	Identify technology hurdles, which must be over come to advance the technology, general comments about the technology, other observations and recommendations.	Sum

EXHIBIT 10.2 Phase 1 Feasibility Review Table

Value rating for each category can be different for each individual performing the valuation. It is best to rate without emotional considerations. Removing favoritism and personal bias for a technology is not always easy for humans to do.

This phase has a possible maximum rating of 25. Every organization will identify acceptable levels an emerging technology must achieve before allocating resources to perform the task required in phase two. Some technology sectors can have numerous emerging technologies which achieve ratings of 20. Determining an acceptable ratings level will depend upon several factors such as the amount of available resources. Increasing the ratings level can ease burdens placed upon an organization and help maintain strategic focus. Conditions can be adjusted such as placing an increased weighting on preferred parameters.

Hold/Stop

Transitioning between the phases requires a review to be undertaken. During a phase review, the technology undergoes an analysis of the results obtained which are documented in the Phase 1 assessment table. There are numerous factors, which determine whether efforts on the technology should be stopped, placed on hold or proceed to the next phase. Some examples for placing a technology on hold can be; the technology does not meet advertised/hyped expectations, patent issues, not in alignment with current corporate roadmaps, knowledge of a superior technology is emerging, or the overall rating is not sufficient. In the event the technology rates relatively well in all parameters, a score card model aides in the decision process especially when comparing multiple technologies or projects. Each organization will develop acceptable rating limits before pursuing emerging technologies and committing resources.

PHASE 2 – Value Research

PHASE 2
Value Research

- Application Analysis
  - ✓ Market/Trends
  - ✓ Technical/Trends
- Value Chain Analysis
  - ✓ Technical Value
  - ✓ Production Value
- Financial
- Risk Identification
- Portfolio Position
- Intellectual Property Position
- Business Design (Preliminary)
- Summary Statement

This phase is where the technology's value proposition is determined. The process moves from a scorecard process to a checklist format generated to ensure accurate data is collected and critical aspects are addressed. Market studies for both established technologies and emerging technologies are performed and early considerations for business models will begin to evolve. Data collected during this phase can be extensive. A brief description of each element with a review of the decision matrix will be discussed. Note that due to industry blocking strategies, funding, technical hurtles, and other issues, this phase can be very fluidic in nature. As time progresses, new technologies can emerge or technological milestones may not be realized, markets and corporations can change directions, loose interest, or form alliances resulting in dramatic changes or shifts in the industry.

Application Analysis

Understanding the fundamental function of the technology will aid in the identification of other emerging and adopted technologies, which provide the same or similar function. These technologies may, or may not, be equivalent but they must be acknowledged and understood. What technologies are currently being used by the industry? To what sector(s) is the technology migrating? Observing industry leaders and their technology migrations will aid in providing insight to answering these questions. Where is adoption most likely occurring for which application? What collaborations are required to enhance the adoption rate? Which companies are best positioned to be a strategic collaborator?

A SWOT (Strengths Weaknesses Opportunities & Threats) analysis from a technical perspective is to be performed on all emerging, competing, and adopted technologies. Investors, stakeholders, industry specialists, and competitors will challenge one's knowledge. There can be attempts to "be-little" the technology. Statements from the industry can be made that are designed to diminish its viability. These statements will utilize both passive aggressive techniques and random arbitrary comments, which may or may not have merit. These must be expected and if a thorough SWOT analysis has been performed, one will have information beyond the hype and the ability to adequately address such issues as they occur.

Value Chain Analysis

Analysis of the value chain will demonstrate where the emerging technology fits within existing applications and most important, how it can be differentiated. Comparing advantages and disadvantages between similar technologies will clarify the emerging technology's value proposition and its value to the organization (internal and external). The emerging technology should be mapped and compared to existing technologies to communicate a clear message of its value proposition. Tabulate differentiators, technical, financial, and "use" specifications, against competing technologies. Similar to the Application Analysis section a SWOT analysis is to be performed from the Value chain perspective. The SWOT analysis for both sections can be performed in a single iteration, on the condition that it includes perspectives of both Value Chain and Applications.

For an emerging technology to be adopted it also needs to provide a cost justification. Can the technology be manufactured and, if so, what are the operational costs? This is to be a detailed analysis, not an estimate. If the emerging technology is attempting to displace an adopted technology, what impact does it have on the adopted process? This analysis is to include the actual process to manufacture the technology and the improvements upon the utilization of the technology in its application. This will provide numbers for cost reduction in materials, capital equipment, time, facility, CO_2, environment impact, personnel, etc.

Financial

During the initial assessment of Phase 2, financial revenue calculations and projections can be difficult to determine. There simply is not enough reliable information available to provide any realistic projection. Especially in terms of projected market share. Whether to use conservative numbers or aggressive numbers depends upon several factors such as organizational effectiveness, history of execution, industry reputation, business model, experience of management team, etc.

It is common for organizations to receive an over whelming number of requests to invest in opportunities. Investments are designed to stimulate growth. Growth can be obtained organically or through an alliance, merger, or acquisition. Decisions have to be made that will prioritize these opportunities. A commonly utilized investment decision tool is the Internal Rate of Return (IRR) or the more recently accepted Return On Investment Capital (ROIC). From a financial perspective the investment alternative with the highest ROIC is preferred. Every company will set an acceptable threshold for an ROIC. When approaching an investor or a tier 1 company it should be understood that they constantly receive and evaluate requests. Often the company with the emerging technology will assume there is immediate room in budgets to accommodate their technology, or that the tier 1 company is not developing an alternative and comparable solution. This section estimates the investment required for subsequent phases, which may include development costs for the technology to be developed internally or externally. Estimating the required investment for subsequent phases requires knowledge of the business strategy, maturity of the technology, and assumes critical milestones will be achieved within projected time frames. Early projections into the estimated impact to product costs can also be determined (i.e. production cost impacts). The following identifies common cost impacts:

1. Technology Impact: estimating the effect the emerging technology will have on products and/or processes from an operational perspective.
2. Development cost: the estimated resources required for research and development, including IP related expenses. This can include such items as building development laboratories.
3. Market, and technology assessment costs: Market and assessment costs are items such as attending conferences, purchasing market reports, evaluating competing technologies, building collaborative relationships, etc.

Managing the process and gathering information to make informed decisions requires resources. The goal is to have the most current and accurate information available.

Risk Identification

Throughout the development of the emerging technology, associated risks will change. Additional risks will be added, some will diminish and some will be significantly altered. Utilization of FEMA (Failure Effects Modes Analysis) and risk profiling techniques for impacts to the technology will be used.

Risk profiling includes identifying sources of risk arising from the market, government regulations, technology, and organizational capabilities. These risks can be listed in a Risk Matrix (Exhibit 4.16) with a significance value and proposed mitigation assigned to each. The identified risks will change or grow to accommodate the emerging technology.

Portfolio Position

Is the technology or application in alignment with the target adopter's technology, brand, and/or values? Successful corporations utilize the element of portfolio position when determining if they should adopt a technology. Portfolio position is another aspect to be considered when evaluating which organization is best positioned to be a strategic collaborator. Initial concepts of where and how to position the technology begin to evolve when understanding competitor and targeted collaborator portfolios.

Intellectual Property Position

What does the intellectual property (IP) map look like? Are there blocking patents? Does the technology or application require the use of patents not currently controlled? Which patents are critical? Can new patents be filed? Are patents exploited throughout the supply chain? This section reviews the patent portfolio required to move a technology from the core IP to the market as described in Exhibit 6.1. Occasionally conflicting patents will exist and will be discovered as a result of a thorough patent search or by disclosure from the market.

Decisions on strategy are to be discussed. Recalling the Taiwan example where Taiwan P.R.C. is not a member of any international patent treaty. What is the strategy for Taiwan? Should the technology be patented in Taiwan? Everywhere except Taiwan? Sell or license Know-How to a targeted organization within Taiwan? The market is global and all strategic planning efforts are to entail a global perspective.

Sector	No. Filed	Licensed & Limitation(s)
Core IP	Patent No.	Company 1, Company 2, etc.
Material Processing		
Manufacturing Process		
Business Design		
Application		

EXHIBIT 10.3 Phase 2 Intellectual Property Assessment

Business Design

Analysis of the technology and its value chains will generate early indicators of an initial business design. This phase is when high-level business strategies can be drafted and proposed. The final business design determined in Phase 4 may not reflect any aspect of the original business designs identified in this phase. This business design is primarily used as a starting point of discussion. It is imperative to follow a standard process for developing a business strategy, ensuring all aspects are properly addressed. At this phase in the evaluation process only a high level approach to a business design is required.

Summary Statement

A brief summary statement provided by the analyzing individual or team, which includes a professional opinion of the technology and applications needs to be provided. Data analyzed and collected are to be summarized in a table, Exhibit 10.4. As projects and opportunities are reviewed over time, this summary statement will provide insight into the next phase's evaluation. Holding a phase review completes the phase. A team of decision makers is assembled and they perform an evaluation of the technology and associated phase contents. As a result of the phase review, required milestones for the technology and application will be identified. The identification of milestones is independent of the Go, or Stop/Hold decision process. If the technology is authorized to proceed to the next phase, milestones are identified and documented. These milestones can be technological or business related. If the technology receives a Stop/Hold status, supporting documentation for the decision will identify what has to occur for the technology/opportunity to become viable again.

Phase 2 pulls together a collective of information about the technology, inclusive of currently adopted technologies and other competing emerging technologies. The technology's position and value are clearly documented. The same occurs for the application. Where has the market been and where is it going? What role can this emerging technology play as the market evolves? Is everything in place as required for market protection and exploitation?

Phase 2 – Value Research		
Parameter	**Comments**	**Champion**
Application Analysis	Summary Comment by Project Team	Name
• ***Market Trends***	Summary Comment by Project Team	Name
• ***Technical Trends***	Summary Comment by Project Team	Name
Value Chain Analysis	Summary Comment by Project Team	Name
• ***Technical Value***	Summary Comment by Project Team	Name
• ***Production Value***	Summary Comment by Project Team	Name
Financial	Summary Comment by Project Team	Name
Risk Identification	Summary Comment by Project Team	Name
Portfolio Position	Summary Comment by Project Team	Name
IP Position	Summary Comment by Project Team	Name
Business Design	Summary Comment by Project Team	Name
Summary Statement	Management Review & Summary	Name

EXHIBIT 10.4 Phase 2 Value Research Review Table

PHASE 3 – Quick Test Market

This phase leverages the data from the previous phases to provide a sanity check on the direction of strategy and business design. Collaborator's portfolios and technology migrations have been identified. A clearly defined value proposition the emerging technology brings to the collaborator has been defined. Additionally the collaborator's role in achieving application adoption has been defined and the information to be disclosed is clearly identified to the management team.
This phase is the refinement phase. It is imperative to build and target the technology to what customers want. This phase will require communications with potential customers.

Some emerging technologies will have milestones to overcome prior to full adoption, which will require collaborators and a strong industry interest. Defining the target product could take significant time and management of industry perception. It will require the building of relationships and collaborations to define an acceptable product specification and increasing over all industry interest (i.e. managing the hype).

PHASE 3
Quick Test Market

- Collaborators
  - ✓ Identify
  - ✓ Define Value
  - ✓ Disclosure
- NDA, JDA, JVA
- Intellectual Property Position
- Risk Management
- Portfolio Position
- Define Opportunities & Requirements
- Analyze Results
- Business Design Review
- Summary Statement

Decisions on the types of agreements and the timing of such agreements are planned. These are tested through the execution of NDAs. Does the collaborator understand the technology and the value proposition? Often, the manager of an emerging technology will need to present current market and technology conditions, which are assumed to be understood. With a thorough analysis performed in Phase 2, the manager of the emerging technology will be prepared and should be prepared to present such information.

Cultural issues can become apparent during this phase. Communications, terminology, and application issues, if they exist, will begin to appear and will need to be managed. Establishment of collaborations demonstrates achievement and validation of strategic direction. It provides confidence to stakeholders by initiating industry interest.

Phase 3 will also identify resources required to effectively implement the strategy. Resources will include such items as personnel, equipment, facilities, and time.

Technology does not stand still. The industry must be continuously monitored for emerging threats and opportunities. Risk will be continuously managed and the IP portfolio reviewed and updated. The completion of Phase 3 will be performed after communication with identified collaborators and the remaining list has been reviewed. This phase is commonly in operation while Phase 2 and Phase 4 are being implemented. It is referred to as "Quick Test Market". An extensive amount of time is not expected to be spent on this phase.

In the event collaborators or the industry does not indicate interest, then a review of the data in Phase 2 is required. Along with updating Risk, IP, Portfolio Position, and Opportunities, the refined business strategy direction will be documented in the Phase 3 review. A review of the phase is performed and summarized as in Exhibit 10.5.

Phase 3 – Quick Test Market		
Parameter	**Comments**	**Champion**
Collaborators	Summary Comment by Project Team	Name
NDA, JDA, JVA	Summary Comment by Project Team	Name
Intellectual Property	Summary Comment by Project Team	Name
Risk Management	Summary Comment by Project Team	Name
Portfolio Position	Summary Comment by Project Team	Name
Opportunities & Requirements	Summary Comment by Project Team	Name
Analyze Results	Summary Comment by Project Team	Name
Business Design Review	Summary Comment by Project Team	Name
Summary	Management Review & Summary	Name

EXHIBIT 10.5 Phase 3 Quick Test Market Review Table

PHASE 4 – Create Action Plan

The previous phases carried several repeated categories critical the management of emerging technologies. These items were Risk, IP, Applications, and Collaborators. Each of these aspects can rapidly change, altering the strategic direction of the business. By following a review process, ensuring that each item is addressed, revisited, and not over looked, provides an effective communication tool to stakeholders. The process allows visibility into decisions, decision processes, and validation of strategic direction.

Phase 4 Clarifies how to move forward. A scope statement is created and serves as a reminder to direction, similar to a mission statement. Personnel will be defined. Individuals with critical skills such as legal, with extensive experience in international agreements, IP licensing, collaborations, Joint Ventures, outsourcing agreements, etc. will be identified. The executive team will require in-depth knowledge of, and experience with business development for international markets and technologies. The team will require technical champions with clear focus and knowledge of their role to achieve the corporate goals. As mentioned previously, this book is focused on the management of emerging technologies, not products. Emerging technologies find their way into consumer based products, which provide enhancements to value propositions.

PHASE 4
Create Action Plan

- Scope Statement
- Define Resources/Team
- Risk Management
- NDA, JDA, JVA
- Budget (Estimate Time & Costs)
- Management Plan
- Communication Plan
- Finalize Business Design
- Create Action Plan
- Project Plan & Approval
- Kick-Off Meeting

Additional resources will be required to provide technical services, with skills to overcome technical milestones and help enable the adoption of the technology by collaborators/customers. A review of the resources required to support the business strategy will be outlined in the *budget*. Critical items are: How much is required? How is it going to be spent? And, how long before revenue begins? A realistic budget based upon extensive data, can be derived, by the time Phase 4 begins.

The *management plan* defines the operational planning for the business strategy. What technology will be outsourced? How will differential advantage/control be maintained? Where is the preferred operational location? What is the corporate management structure? Whereas the *Communication Plan* identifies how and in what form information is disseminated to stakeholders and employees. It can also identify operational metrics for monitoring the business and technology metrics for monitoring the management of milestones and industry trends. These plans are drafted to clearly communicate initial direction and provide stakeholders the opportunity to make recommendations for a unilateral understanding. Once agreed upon, the management team uses the plans as initial direction. As markets and industries change, the business strategy will adjust to these changes. It is not recommended that the team update these documents. The team should be allowed to pursue the directions required to achieve success.

During the Phase 4 review, it is a good practice to pause and perform a sanity check. Review the direction from a simplified perspective. For example in the case of RFID, industry has claimed that the RFID market will "Take-off" when the price per tag drops to less than US$0.06. Expecting the cost to purchase the chips, program them, validate the programming, print an antenna, assemble the chip to the antenna and to the package, 100% reliably, for a cost of less than US$0.06 each? Notice that capital equipment, G&A, and profit were omitted? Is it really a good idea to spend the capital to manufacture a product into a commodity market? Is it a good idea to carry all that CAPEX to manufacture billions of tags, supply them to the market, and expect a profit? This is similar to building a semiconductor factory with the idea that one will successfully compete in the memory chip market. Profitable business models do exist for RFID. It is important is to perform a sanity check on the direction and business strategy. Sometimes investors and technology managers can get caught up in hype and over look the basics.

Phase 4 – Create Action Plan		
Parameter	**Comments**	**Champion**
Scope Statement	Define Business Scope	Name
Define Resources	Assign Key Business Resources	Name
Risk Management	Review & Monitor Risks	Name
NDA, JDA, JVA	Review Status	Name
Budget	Estimate Budget	Name
Management Plan	Management Plan	Name
Communication Plan	Communication Plan	Name
Finalize Business Design	Outline & Define Business Strategy	Name
Create Action Plan	How? When? Who? Where? What?	Name
Summary	Phase Review, Project Plan Approval & Kick-Off Meeting	Name

EXHIBIT 10.6 Phase 4 Create Action Plan Review Table

PHASE 5 – Implementation

PHASE 5
Implementation

- Execute the Plan
- Information Dissemination
- Risk Management
- Team & Stakeholder Development
- Scope Verification
- Progress Meetings

Execute the Plan, build and monitor the business, drive to adoption. Monitor market technology trends and emerging threats. Utilize the table in Exhibit 10.7 to establish acceptable formats for reports. Disseminate information in accordance to the communication plan to stakeholders and employees. Risk management will continue to be monitored and mitigated. Team and stakeholder management is an ongoing task.

Hire the best available resources. Place a champion at the top of the organization. If team members don't have the skills, train them, utilize executive training programs from leading universities. Don't start "Empire Building". Keep the resources realistic and balanced. Outsource where possible. Remind employees of the value proposition the organization provides to the customer, and their role in helping the organization achieve its goals.

Phase 5 – Implementation		
Parameter	**Comments**	**Champion**
Execute The Plan	Verified by Reports/Track Scope	Name
Information Dissemination	How? What Information? To Whom? What form or Format? Metrics?	Name
Risk Management	Ongoing Monitoring of Risks	Name
Team & Stakeholder Development	Ongoing team Development/Building & Stakeholder Management/	Name
Scope Verification	Review & Update Scope, Modify as Needed	Name
Progress Meetings	As required (Daily/Weekly/Monthly)	Name
Summary	Track and Monitor Metrics	Name

EXHIBIT 10.7 Phase 5 Implementation Review Table

As the plan unfolds, verify the scope is in alignment with the corporate strategy and goals. Allow the organization to adjust to changing market conditions. Hold regularly scheduled meetings to review and monitor progess at a high level.

Monitor executives and managers with respect to corporate culture and communications. Be cautious of distractors to research and development, and to the corporate goals. Utilize tools outlined in the previous chapters to develop metrics and monitor the business. Ensure R&D is both Value Proposition to the customer, and IP generation focused.

Summary

Managing an emerging technology to achieving marketplace success is a path filled with failures and successes, both major and minor in nature. The comprehensive knowledge required is not held by one person and is best served by formulating a highly skilled team. Success will be determined not only by adoption but how long it takes to achieve adoption. Time for adoption can be reduced by following the process. The process is designed to identify and manage the critical issues an emerging technology must undergo. The phases provide the opportunity to properly investigate and provide a solution to the critical issues. It ensures management will investigate and make informed decisions on items, which are commonly overlooked. A review of Phases helps management evaluate where they are and where they are going. It keeps the customer's value proposition on track and can identify distracters.

It is often during early phases where investors of emerging technologies want to see established collaborations with tier one or tier two market leaders. Development of such collaborations demonstrates industry interest in the technology and provides confidence in their investment. At the early stage of evolution, collaborations are not common. This is when the individual assigned the task of managing the innovation holds numerous meetings with the key objective of obtaining focus for application and adoption. Blue-sky ideas and concepts need to be listed, but shelved for future research. The immediate need is a quick win. Identification of what milestones need to be overcome for use in a specific application is the objective of the innovation manager. This information is to be used for directing research and development in a focused effort. Only after demonstrating proof of concept and achieving milestones will an industry leader demonstrate interest. Demonstrating feasibility of technology is only the first step. Subsequent milestones will often emerge which will determine if the technology can overcome known secondary issues. The initial milestones are often asking for proof of concept, while secondary milestones will include processing and design issues for application. Every research and development department has a list of projects under development, with a defined budget and allocated resources. For an emerging technology to be adopted in a collaborative effort from an industry leader, current programs in development will have to be displaced or delayed. Thus the emerging technology will require a compelling reason to justify such a move on behalf of the collaborator.

Early preparation of financial forecasts should be performed on existing markets. However, this data can only be used as reference material. It is not easy to predict future sales of emerging technologies, especially if the technology can leapfrog or disrupt current technologies.

Chapter 11

Poly-OLED Case Study

One of the most exciting developments in the display industry in the past couple decades has been the discovery and development of Polymer Organic Light Emitting Diodes (Poly-OLEDs). It is a plastic material used in the manufacturing of electronic displays. It emits light as a function of its electrical operation and does not require additional elements such as backlights and color filters as does liquid crystal displays (LCDs). Poly-OLED technology is very energy efficient and lends itself to the creation of ultra-thin lighting displays that will operate at lower voltages.

The following case study demonstrates how managing an emerging technology not only requires extensive technology and market knowledge but also timely management of market dynamics. It will also demonstrate with an example of market threads how they converge while driving to successful adoption. Supply chains have to be created and can completely change within a short time frame. Technology collaborators constantly strategize, attempting to position themselves with increased market control. Constant monitoring of the market and the ability to act quickly are crucial for an emerging technology to attain an advantageous position.

The Invention

On February 9th, 1989, Dr. Jeremy Burroughes was studying the 'Franz-Keldysh' effect on conjugated polymers under the supervision of Dr. Sir Richard Friend in the Cavendish Laboratory, Cambridge University (UK). The Franz-Keldysh effect is a change in optical absorption when an electric field is applied. While setting up the experiment, which involved wiring 100 micron gold wires to a cryostat, he noticed another experiment on a similar device about five feet away. Glancing over at the other device, Dr. Burroughes noticed it was emitting green light. Initially he thought it was an accidental reflection from the green phosphor in the computer monitor. Upon moving closer to the screen, he realized the light moved across the screen, quickly dispelling the idea that the computer monitor was the source of the light. As he turned the voltage source off on that experiment, the light went away. Believing this to be an opportunity, he copied the contact arrangement of this experiment and connected his Franz-Keldysh device in exactly the same way. He slowly increased the voltage on his device until light was emitted. At that moment, he discovered a method to emit light from a polymer (plastic). This was the birth of Poly-OLED.

Immediately, Dr. Burroughes could envision the potential value of this technology in numerous future applications. He and his colleagues first approached the university but they quickly discovered the university had no means of filing patents. So they located a patent agent and filed the patent for themselves.

The first discovery device and experiment for Poly-OLED managed to emit light for three days, a very limited lifetime. After two years of further development, the first working displays were developed and were comprised of simple 3x5 pixel arrays. To bring the technology to market, it required the establishment of a corporation and investor funding. With seed capital funding, they rolled it into a company named Cambridge Display Technologies (CDT) in 1992.

Five years had past with limited growth, due to some shareholders unwillingness to be diluted, which limited their access to greatly needed additional funding. Then in 1997, CDT received an investment of $10 million from a financial group headed by Lord Young of Grafham, former UK Secretary of State for Trade and Industry. Later that year, Intel Capital also invested in CDT.

These investments allowed them to develop a fully operational prototype. The first device emitted a single pixel, which could be used in a Light Emitting Diode (LED) application or assembled in an array for a display application. Fundamental research focused on chemistry and driving circuits designed to improve switching frequencies, brightness, illumination lifetimes, efficiencies, and color spectrums.

As with all emerging technologies, commercialization requires customer adoption of the technology. Therefore, identification of applications with achievable specifications is critical. Existing market analysis for electronic displays identified several sectors with significant size and growth potential for which the technology could be exploited. Major applications included electronic display panels (mobile phones, PDAs, etc.), photovoltaics (solar cells) and sensors. CDT decided to pursue display market applications as Poly-OLED offers significant value in materials reduction, enhanced viewing angles, and lower power.

Market Thread One – Epson ink jet printing IP

Meanwhile, news of this Poly-OLED technology had spread to the other side of the globe. Dr. Tatsuya Shimoda of Seiko Epson Corporation (Japan) learned of the material's viscosity and commenced to investigate if it could be ink jet printed. Utilizing a precursor already synthesized by Sadao Kabe of Seiko Epson, combined with Epson's extensive knowledge of ink jet deposition processes in ink jet printing of color filters, Hiroshi Kiguchi of Seiko Epson created the first ink jet printed Poly-OLED device. In September of 1996, Dr. Shimoda visited CDT to investigate the possibility of collaboration. He showed them a photo which demonstrated the possibility of Poly-OLED patterning, but he did not disclose how it was created. Subsequently, on November 25th, 1996 Dr. Shimoda applied for and received a patent for creating Poly-OLED devices using the ink jet printing method.

Using the process of ink jet printing to deposit material enables the ability to manufacture displays without the size restrictions of vapor deposition or photolithographic processes. The spin coating process wastes 94% or more of the Poly-OLED material while ink jet printing wastes only 3 to 5%. With limited industry exposure and market adoption, early material cost projections were US$2000 per gram. With cost estimates at such a high value, the decision to which manufacturing method to implement becomes obvious. Ink jet printing rapidly became the preferred method of manufacturing.

CDT's Critical response

In 1998, Seiko Epson was the only company demonstrating ink jet printed displays. CDT recognized early that ink jet printing would prove to be a critical enabling technology and secured a cross license agreement with Epson in 1999. This gave CDT the right to sublicense the IP and promote this technology. Ink jet printing of Poly-OLED is viewed as a critical patent enabling the industry to develop Poly-OLED display products.

Other deposition technologies such as photolithography and chemical vapor deposition (CVD) were well established but not viewed as the preferred methods of manufacturing for Poly-OLED. Monitoring industry developments and relationships of ink jet printing Poly-OLED was critical to CDT.

> *"Dr. Shimoda is like the Leonardo da Vinci of printed electronics. He may not live to see all his ideas come true, but he had all the basic concepts drawn out well before anyone else. I can honestly say I came up with the idea of IJ printing electronics independently, it was disappointing when I found out I wasn't the first"* – Chuck Edwards.

Dr. Shimoda foresaw the significant impact ink jet printing would have on the future of manufacturing. Traditionally, technology evolves from analog to digital. Audio transitioned from tape to compact disks (CD), movies from the VCR (Video Cassette Recorder) to DVD (Digital Video Disc), public television from analog antennas to TiVo™, and the mobile phone from the large analog brick sized phones of the '80s to the small digital handsets of today. The next digital revolution is manufacturing. Ink jet printing is a digital fabrication process, which can be used to create products. A few years prior to the disclosure of Seiko Epson's ink jet printed Poly-OLED device in 1996, Seiko Epson had already invested heavily in research and development of applications for ink jet printing of electronics.

By 2000, Seiko Epson and CDT had demonstrated the world's first ink jet printed, full color, active matrix, Poly-OLED display. Seiko Epson had also developed an ink jet production line to manufacture color filters for their display products. Leveraging their advanced research on Poly-OLED, in-depth display knowledge and usage of ink jet printing in a manufacturing environment, Seiko Epson decided to investigate ways to capitalize on the emerging market of ink jet printing Poly-OLED displays.

Exhibit 11.1 presents a graphical representation of the Poly-OLED supply chain logistics. It identifies the core competencies Seiko Epson brings to the production process. Through collaboration proposed by CDT, Seiko Epson architected a business strategy around their core competences and investigated the prospect of selling ink jet printing equipment to display manufacturers under the trade name of Polyink. Under Polyink, a display manufacturer would purchase the equipment and pay a per sheet "use" fee throughout the life of the product, while Seiko Epson would warrantee the production yields.

To achieve this Seiko Epson required the ability to control all aspects of the production process. This included the purchase of materials such as Poly-OLED. Display manufacturers knew that ink jet printing had rapidly developed into the preferred method to manufacture Poly-OLED displays. Other than creating their own equipment, which some tried, display manufacturers only had Litrex or Polyink to choose from. Selecting Litrex gave the display manufacturers the ability to develop their own processes and acquire the Poly-OLED material directly from the material suppliers. Some display manufacturers viewed this as a strategic advantage opportunity to develop ink jet printing expertise in-house. While others understood that ink jet printing is far more complex than it appears. The desktop ink jet printer left the impression that the printing process is extremely easy and any qualified manufacturing engineer could duplicate it. The information not on the surface of this impression is the extensive amount of research in chemistries for the surface of the substrate and the chemicals being deposited. Additionally and equally important is the required post processing expertise required to cure the inks.

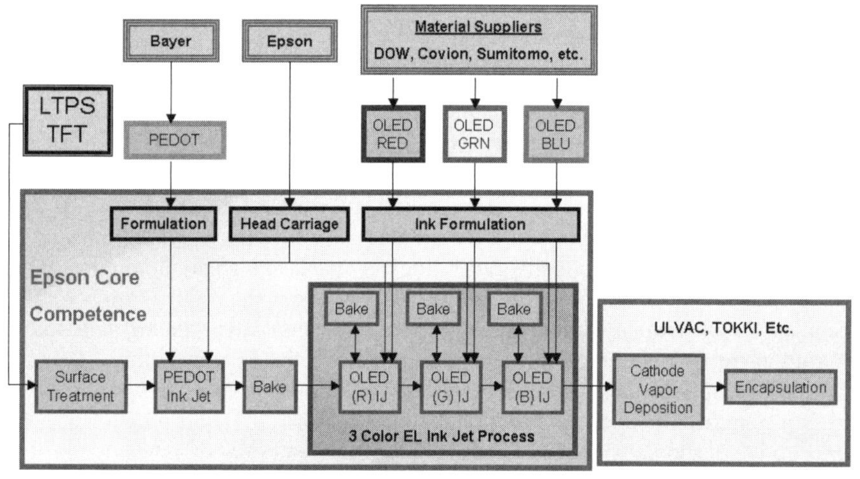

EXHIBIT 11.1 Supply Chain Logistics for Poly-OLED

Courtesy: Seiko Epson Corporation

CDT's involvement with Polyink ensured technical collaboration with Epson and display manufacturers. It also created a turn-key supply chain which included material, processing, and device manufacturing for the market.

Market Thread two – Litrex Corporation

Across the continent in 1998, Gretag Imaging AG (Switzerland) acquired Raster Graphics (USA) and their ink jet technology business Litrex Corporation.

Gretag Imaging made most of their revenues from film processing equipment sold to Walgreens and other 1-hour photo shops. When 1-hour photo came out, it was big. All the stores bought them not to make profit on film processing but because it guaranteed customers would either spend 1 hour wandering around the store, or they made two trips to the store. Either way it increased sales. Stores were leveraging the consumer's tendency to impulse buy when getting their film developed.

The advent of digital cameras meant the photo developing equipment would have to go digital as well. Gretag reacted quickly. They recognized that digital printing for film development was a natural solution to the digital revolution so they acquired Raster Graphics. As part of the acquisition, Raster Graphics' ink jet technology business, Litrex Corporation, was acquired for US$6 million.

Litrex specialized in the development and sale of high precision ink jet machines for development and production of electronic applications. Headed by Chuck Edwards, there were only 6 individuals on the team, they had the 80L design and the 140 concepts started. They had also filed fundamental patents. Litrex was the first company in the world to offer for sale a commercial high precision ink jet system designed specifically to print displays.

Litrex did not provide ink jet print head, processing, or display technologies like Seiko Epson. Instead they purchase ink jet print heads from manufacturers such as Xaar and Dimatix (formerly Spectra). Display manufacturers could obtain a high precision ink-jet printing machine from Litrex. The accuracy of the Litrex machine was far better than most could quickly create but not as accurate as Seiko Epson's. The Litrex machine came fully configured and without the extensive ink jet printing knowledge base. Development of the process was left to the display manufacturer. Many companies tried to acquire Seiko Epson's ink jet technology independent of Polyink, but Seiko Epson viewed their technology as a key-enabler providing them a strategic advantage and would not sell it. Over time other companies developed industrial quality ink jet printing machines such as Creo (Kodak) and iTi Corporation.

In 2001 Gretag Imaging was having financial problems. They informed the industry that some of their business units were up for acquisition. With only two viable ink-jet printing solutions on the market, one being Seiko Epson who clearly was not making their technology readily available, and the other was Litrex. Learning of Gretag's financial problems, Covion, a Poly-OLED material manufacturer, looked into securing Litrex for the purpose of strategic control. This would provide Covion strategic market control by enabling display manufacturers with Poly-OLED ink jet printing capabilities to only those who purchased Covion materials. This meant CDT was about to loose control of the supply chain and one of the critical enabling technologies.

CDT's critical response

In May of 2002, to prevent Covion from acquiring the only commercially available ink jet printing technology, CDT quickly acquired Litrex from Gretag Imaging for US$10 million. Along with Litrex, CDT obtained the rights to use Raster Graphics patents for ink jet printing electronic applications. This acquisition yielded a US$4 million profit for Gretag Imaging and helped CDT maintain control of the Poly-OLED technology supply chain.

This move in conjunction with the ability to license ink jet printing of Poly-OLED from Seiko Epson allowed CDT control of the technology within the supply chain. Display manufacturers now had the option of developing their own process with any Poly-OLED material or signing up for the Polyink solution.

Market thread three – supply chain development

Moving the technology beyond the lab into the market requires identifying its value proposition within the targeted application. Poly-OLED technology required the development of materials, processes, and adopters. Since the Poly-OLED technology was emerging, there was no established infrastructure to leverage. CDT had to build the value chain. A license to manufacture and sell displays was granted to Uniax Corporation (On March 16, 2000 Uniax was acquired by DuPont Chemicals), and a similar license was granted to Philips. By granting these two licenses, CDT built a simple value chain and increased value to shareholders.

As CDT published material performance data they began to notice inconsistencies in data reported from collaborators. Each display manufacturer was building and testing their independently developed devices. Test methods and device structures were not consistent resulting in different data. This added confusion to the market and did little for the reputation of Poly-OLED material. CDT attempted to correct this inconsistency by establishing standard testing procedures and device structure. Several display and material manufacturers were invited to a meeting held at the Society of Information Displays (SID). The significant take-away from the meeting was that each attending member wanted to collaborate but they also wanted to protect any opportunities to develop IP. This meant that each member relied on CDT to perform all disclosures as the members withheld information which could enhance the advancement of Poly-OLED technology.

CDT's critical response

CDT realized that without the ability to build their own displays, they were allowing display manufacturers the opportunity to control significant process IP. CDT needed display device processing expertise and decided to build a small, yet comprehensive, display production line in Godmanchester UK. With this new production facility CDT was able to enhance productivity with their collaborations, expand opportunities for IP, and advance the Poly-OLED technology.

Life cycle

It was not until 2002 that OLED became increasingly adopted in consumer products. Philips launched one of the first consumer products containing OLED technology in their innovative electric shaver with an electronic display, which was featured in the James Bond movie, "Die Another Day". More than 15 years after the discovery, cell phones and PDA's became increasingly available with OLED technology. Envisioning future applications, the James Bond movie covered a car with OLED for camouflage; it displayed the background of the surroundings on the surface of the vehicle, rendering it invisible to the viewer. In the movie "Minority Report" an electronic newspaper was digitally updated to report a fugitive (Tom Cruise) on the loose. As a technology matures, time and innovation will tell how and where it will be adopted.

The hype for Poly-OLED had been on the up hill climb to the peak of over inflated expectations. By December of 2002 interest in OLED fell to the *Trough of Disillusionment.* Struggling with lifetime issues, encapsulation issues, and secondary processing issues display manufacturers began to realize that the material suppliers still required additional time on research. The technology was premature for the market. These companies stopped announcing successes at conferences and decided to focus on solving the remaining issues surrounding the technology. This silence from the industry resulted in the perception that the industry had moved away from the technology. Investor interest and magazine publishers followed this perceived direction. With investor funding for CDT becoming increasingly difficult to obtain, stakeholder management became the highest priority for CDT. When a technology finds itself in the *Trough of Disillusionment*, the effort required to realize the true value can take a significant amount of time and dedication; coupled with patience.

Only by making difficult decisions was CDT able to survive. CDT modified its business model by abandoning small volume display production at the Godmanchester facility, refocused R&D and laid-off 25% of its resources. In a collaborative effort, the material suppliers and display manufacturers began to focus on discovering the true limits and capabilities of the materials. These efforts migrated the technology onto the *Slope of Enlightenmen*t. Over the subsequent eighteen months, a few display manufacturers demonstrated proof-of-technology in display sizes ranging from 2 to 17 inches. In May of 2004 the hype returned to OLED when Epson announced a breakthrough by demonstrating the world's first 40-inch inkjet printed Poly-OLED television.

Poly-OLED slowly continued to gain market interest and adoption; OSRAM announced products using Poly-OLED displays, Toppan Printing (Japan) announced the building of a ¥1bn pilot line to investigate alternative printing methods, including roll processes, for Poly-OLED display manufacturing, Philips incorporated Poly-OLED displays into their 'Magic Mirror' mobile telephone, and MicroEmissive Displays (MED) produced a display for a 3-in-1 miniature camera and MP3 player.

By July of 2004, CDT was ready to file for an Initial Public Offering (IPO) with the Securities and Exchange Commission. On December 16th they listed on the NASDAQ under the ticket symbol "OLED". CDT's stock price closed on the opening day at US$11.27/share resulting in a market capitalization of US$230 million.

Supply chain disruption

During the period when Philips launched their Magic Mirror product, Poly-OLED material had not scaled up for mass production. There were three primary Poly-OLED suppliers licensed by CDT on the market; DOW Chemical Company, Covion and Sumitomo Chemical. While ramping for mass production DOW Chemical made the decision to halt investments and placed their Poly-OLED technology (Lumation™) up for sale. Lumation™ along with all the associated intellectual property was acquired by Sumitomo Chemical. Covion was also sold to Merck KGaA. These two acquisitions disrupted the material supply, making the production of displays difficult. Philips decided to halt production of the product and sold the Poly-OLED production line to the OTB Group who created OTB Displays with the technology. This material supply interruption also disrupted developments and relationships. New collaborations had to be established while some were re-negotiated.

With an IP licensing business model, CDT needed to continue managing the control of Poly-OLED related IP. Sumitomo's acquisition of Lumation™ occurred in May of 2005. CDT quickly moved to create a joint venture with Sumitomo Chemical that same month. The new company was named Sumation™. Sumitomo Chemical and CDT brought together the largest concentration of Poly-OLED expertise and intellectual property into a single organization.

Final push

Due to the limited lifetimes for the color blue Poly-OLED continued to struggle with obtaining full market adoption. Stock prices began to decline as market interest began to slow and UDC's alternative small molecule material was obtaining design wins.

Fortunately, CDT's joint venture with Sumitomo Chemical (Sumation™) proved to be beneficial. Their combined technologies resulted in achieving record material lifetimes. In April of 2007 market interest was revived when Toshiba Matsushita Display Technology (TMD) demonstrated a 21 inch Poly-OLED television and Sony's repeated announcement of developing a mass production line to manufacture 11 inch OLED televisions. These announcements brought OLED up on the Hype curve again. Display manufacturing companies who waited for someone else to make the first move, began to aggressively start mass production plans.

On July 31st of 2007 with 21.6 million outstanding shares, Sumitomo Chemical announced the acquisition of CDT for US$285 Million.

The "Work" behind the scenes

It took CDT 15 years from the discovery in 1989 to achieve IPO in 2004 and an additional 3 years to be acquired. This achievement was accomplished through diligent market analysis and channel management. As demonstrated in the case study, multiple market threads converged and effective management brought Poly-OLED to the eventual successful sale of the company.

The different phases of CDT's development can be analyzed using the technology evaluation process described in chapter 10. The intention is to demonstrate by example, the types of challenges and adjustments in strategies one must be prepared to manage while attempting to bring an emerging technology to adoption.

Phase 1 - Feasibility

Phase 1 of Feasibility encompassed the period from Eureka in 1989 until the first licensing agreements obtained in 1996. Exhibit 11.2 is how the Phase 1 Feasibility table for CDT might look like it a review had been carried out 1996.

Parameter	Comments	Rating
Description	Light Emitting Polymer	- - -
Maturity/Feasibility	Demonstrated 3x5 Pixel array	3
Application(s)	Displays/Signage, LED, Photovoltaic	5
Differentiation/ Value Proposition	Thinner, potentially more efficient, significant cost reduction (projected), low cost manufacture	4
IP Status	USA: Received 11, Filed 47 JPN: Received 10, Filed 155	5
Portfolio Alignment	Aligns with current strategy and core competence	5
Summary	High confidence level of achieving technology goals, strong IP position. Milestones of RGB Colors and life have to be overcome. Production will require large-scale chemical manufacturers. Business model is IP licensing & display manufacturing.	22

EXHIBIT 11.2 Phase 1 Feasibility summary of CDT's Poly-OLED Technology

- *Maturity/Feasibility:* Maturity of the technology defines the level of risk during the Feasibility period; the technology was limited to monochrome devices due to material lifetime limitations. Only a limited number of laboratory developed prototypes were observed throughout the industry. Devices were being developed in a few laboratories; material suppliers and display manufacturers were only beginning to show interest. With the technology at this level of maturity it can be ranked as a "3".
- *Applications:* On the application end, CDT fared much better. After seven years of diligent work CDT had acquired industry visibility and the beginnings of critical collaborations for material suppliers and display manufacturers. They

demonstrated a 3x5 pixel array with the technology and pursued patent applications. CDT managed to obtain industry collaborations from Uniax and Philips. This gives CDT a high applications rating of '5', with further technology analysis needed.

- *Differentiation/Value Proposition:* Poly-OLED technology was focused on display applications, a large diverse market with significant growth potential. There are several key value differentiators such as; front emissive technology, potential lower manufacturing cost due to fewer components, thinner, and lower power. Early evaluations for CDT's value proposition give it a rating of 4.
- *IP Status:* CDT had received 21 patents around the core technology, material, applications and processing. Other companies with patents or prior art, did not come close to challenging the technology. This strong patent position places them with an IP rating of '5'.
- *Portfolio alignment:* If the Feasibility evaluation had been conducted by a display manufacturer, CDT would receive a rating of '5' due to alignment with corporate objectives focused on display industry applications. If the portfolio alignment is performed by a financial investor, CDT might receive a rating of 3, due to limited relationships with other potential applications' and industry channels, such as in photovoltaic applications.

The researchers within CDT demonstrated a high level of confidence in their ability to overcome performance milestones. They were also well tuned to the value of IP and maintained an IP portfolio which was increasing in number.

In 1995 it was demonstrated that Poly-OLED technology could be used to manufacture photovoltaic devices. Should CDT spend resources to develop photovoltaic devices? The evaluation of Poly-OLED for displays fits within CDT's objectives and strategy, while photovoltaic development did not.

Phase 1 is a quick assessment of the emerging technology and as reviewed, places the Poly-OLED technology in a relatively high rating, 22 out of a possible 25.

PHASE 2 - Value Research

Phase 2 will analyze the Poly-OLED technology from the licensing agreements obtained in 1996 through the end of 2002. Exhibit 11.3 is the summary table for the phase 2 evaluation of Poly-OLED. Contained within the comments column of the table are brief statements as a result from the analysis for each parameter.

Phase 2 – Value Research			
Parameter	**Comments**	**Author**	**Report**
Application Analysis	LEDs & Displays	- - -	Documented
• ***Market Trends***	Lower Power, Thinner, Lower cost	- - -	Documented
• ***Technical Trends***	Increasing LCD, decreasing STN, increasing OLED	- - -	Documented
Value Chain Analysis	- - -	- - -	Documented
• ***Technical Value***	Thinner, lower power, improved viewing angles	- - -	Documented
• ***Production Value***	Fewer components, solution processing, lower CAPEX	- - -	Documented
Financial	Excellent investor support	- - -	Documented
Risk Identification	Life of Blue & encapsulation	- - -	Documented
Portfolio Position	Alignment with Displays	- - -	Documented
IP Position	131 patents, more pending	- - -	Documented
Business Design	IP model	- - -	Documented
Summary Statement	Market indicating cautious adoption, waiting for life of blue repeatability.	- - -	Documented

EXHIBIT 11.3 Phase 2 Value research summary of CDT's Poly-OLED Technology

Application Analysis

Displays have several market sectors such as; mobile phones, PDA, automobile monitor, digital still camera, video camera, amusement, notebook computers, etc. Data for technology migration or trends for each market sector are readily available, examples are in appendix A.

An application analysis studies the market for not only growth sectors, but trends in technology, identification of competing technologies, and the trends of industry leaders. Starting with a fundamental overview of the emerging display technologies, Exhibit 11.4 shows the evolutionary family tree for flat panel displays. It reflects the technology evolutions that lead to the emergence of Poly-OLED.

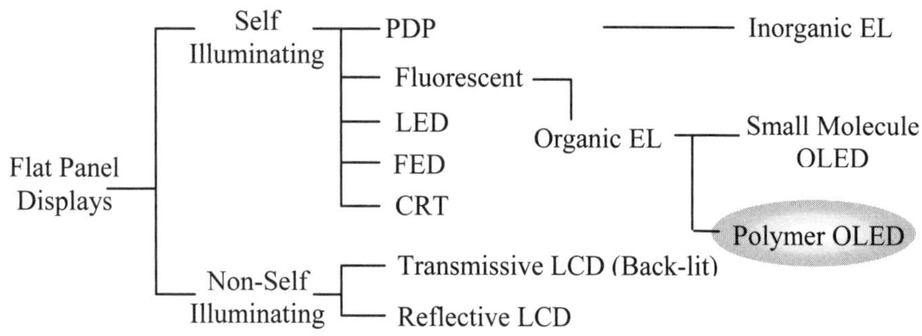

EXHIBIT 11.4 Flat Panel Displays Family Tree

Notice the divergence in the branch for Polymer OLED and Small Molecule OLED. Both are Organic Light Emitting Diode technologies. The difference between them is their chemistry; Polymer OLED (Poly-OLED) is a liquid solution while the competing technology, Small Molecule (SMOLED), is a dry powder. This is very significant especially with regards to processing variations. Liquids can be poured, ink jet printed, spin coated and meniscus coated. While dry powders use CVD and PVD types of manufacturing processes. Both have their respective advantages and disadvantages. However, Poly-OLED has the ability to be processed in large formats, whereas SMOLED has substrate size limitations.

Application Analysis – *Market Trends*

Exhibit 11.5 reflects 2002 conditions for the small/medium display market. It shows the potential market size in terms of revenue and number of units for different segments. It is used as an aid when analyzing which application to target. Further data is also available which identifies the leaders in these markets. As the data indicates, the mobile phone market consumes the most material in this market sector and is of primary interest to CDT.

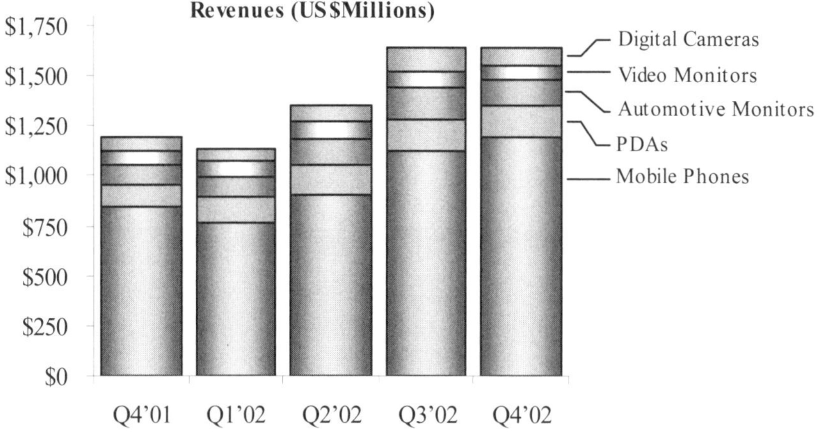

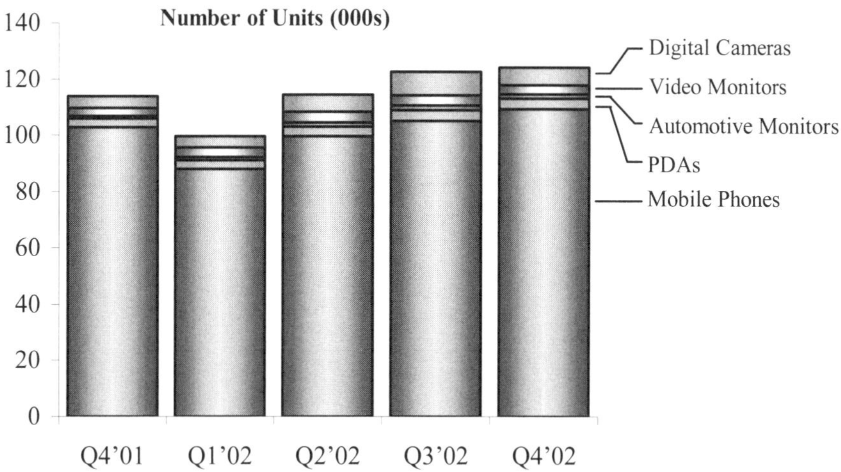

EXHIBIT 11.5 Q4'01 – Q4'02 Small/Medium Display Market Conditions

Source: DisplaySearch, an NPD Company

Identification of industry leaders and in which technologies they are investing will prove to be crucial when establishing collaborations. It is important to strategically place the technology in the path of the market, rather than in direct competition. This is achieved by collaborating with key market leaders and demonstrating technical feasibility. Exhibit 11.6 identifies the top tier display manufacturers with respect to annual shipments. Will the technology receive aggressive support from a lower tier organization or will it receive better support from a tier 1 organization attempting to maintain their market position? These are questions that must be thoroughly evaluated when identifying and targeting strategic collaborators. Tier 1 organizations know their brands and channels. They also understand the efforts required to build them. They tend to invest in technologies which support their channels and have the potential to rapidly launch a new technology into their existing channels.

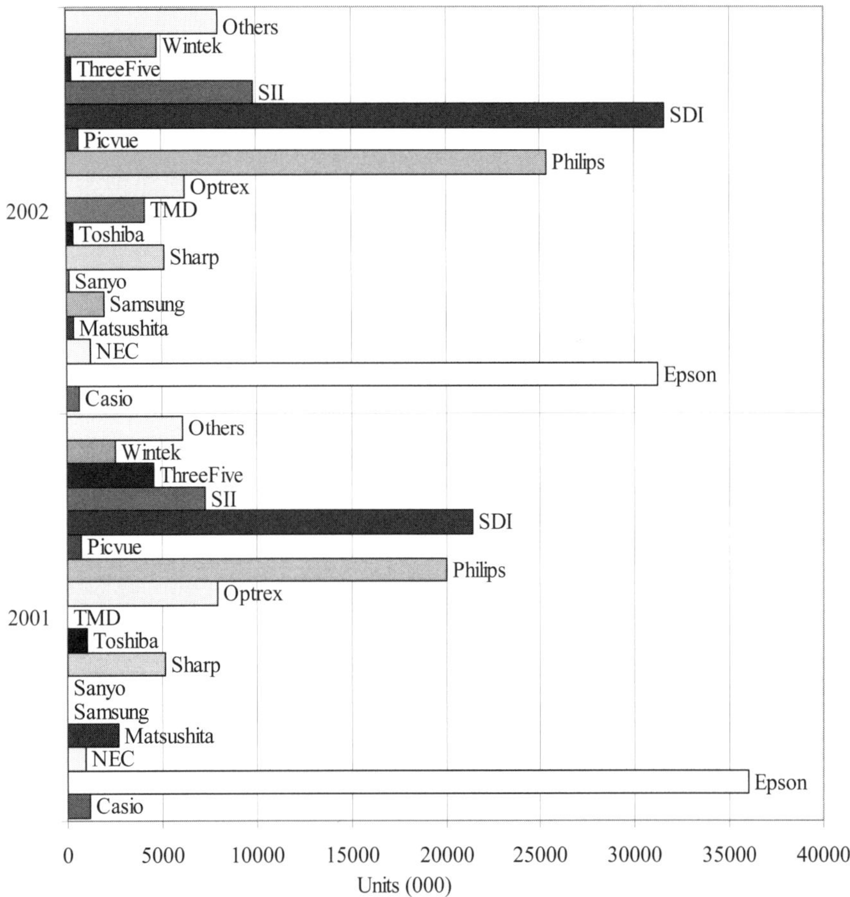

EXHIBIT 11.6 2001 & 2002 Small/Medium Display maker Annual Shipments

Source: DisplaySearch, an NPD Company

In 2001 and 2002 the three market leaders for this segment were Epson, Samsung SDI and Philips. Collaborating with these three leaders would allow access to information related to consumer market pressures and held the potential of rapid adoption. The digital conversion of photography enabled consumers to take and share photos. This resulted in market pressures for mobiles which are smaller, had longer battery life, and high quality displays. Poly-OLED technology could help the mobile phone manufacturers achieve these market demands.

Application Analysis – *Technical Trends*

Exhibit 11.7 lists several display technologies, which compete for market share in numerous applications, each with large market opportunities, for the years 1998 through 2005. The Field Emission Display (FED) technology started to achieve market adoption in 1999 but the numbers of units were too small to indicate a notable percentage. By 2002 the utilization of FED technology returned to the previously insignificant percentages. Stakeholders in OLED technologies did not wish to witness a similar performance.

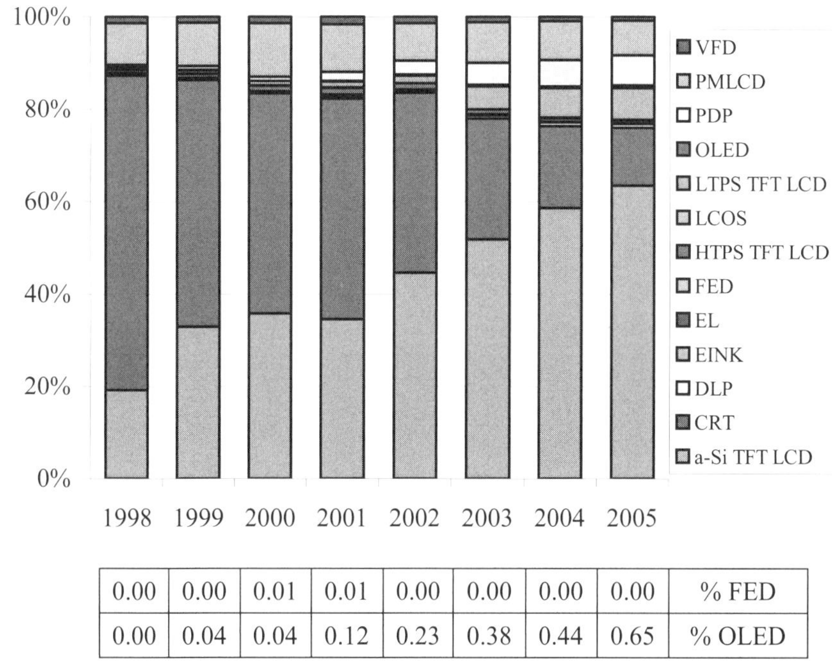

0.00	0.00	0.01	0.01	0.00	0.00	0.00	0.00	% FED
0.00	0.04	0.04	0.12	0.23	0.38	0.44	0.65	% OLED

EXHIBIT 11.7 Shipped Display Technologies as a Percentage of Revenue

Source: Displaysearch, an NPD Company

Additionally, Exhibit 10.6 shows what percentage of the market utilizes the technology. It depicts both technical trends and overall market growth. The decision as to which market sector resources are to be focused dependents upon numerous factors. These factors can include manufacturability, size limits, resolution, color gamut, lifetime, cost, and corporate relationships.

What are other display manufacturers doing? How is their progress? A few industry announcements were made and can be documented. For example; Samsung announced they will introduce OLED displays in late 2003, Philips introduced Poly-OLED in their shaver and announced they will introduce Poly-OLED displays early 2003 in Palm devices. By paying attention to industry announcements and gathering information from established personal relationships (Guanxi). An industry OLED status for 2002 was complied and is presented in Exhibit 11.8.

Manufacturer	**OLED Status**
SDI	Relation with NEC. Disastrous launch. Showing heavy interest in CDT material. Polymer research in Holland.
Philips	Reducing OLED investment. Acquired Litrex Machines for internal ink jet and OLED development.
TMD (joint venture by Toshiba and Matsushita)	Evaluating License conditions from CDT. Favor small molecule. CDT recommends Epson should provide incentive to help TMD make the decision.
Sharp	CDT held 4th meeting in 8 months. Showing strong interest in CDT material. Waiting for lifetime of blue.
ST-LCD	Sitting on 32 inch functional display built using flexographic process. Have shown to only exclusive people. They believe OLED is for large screen applications only, OLED will be the next big screen TV.
Hitachi	Display business is up for sale. Selling LTPS.
Samsung	Very little activity, no budget to install equipment or pilot line. Experimenting with UDC materials
Casio	Polymer focused, very little activity.
Optrex	Taken Kodak License only as a defense move.
Sanyo	Relationship with Kodak, showing strong interest in Polymer OLED. Held two meetings with CDT.
Au Opto	Limited effort, focus on small molecule.
NEC	Joint Venture with Samsung.
Wintek	Small molecule.
DuPont/Ritek	Commercializing small molecule (in production) low yield and quality. Relationship with Dupont beginning to weaken.
Osram	Will introduce cell phone screen in Q1 2003 for a European carrier.
Formosa Plastics	Built production plant, Osram owns 25% of production.
Delta Engineering	Building a big facility (10 million sq-in/month)

EXHIBIT 11.8 OLED Industry Status for 2002

Source: Cambridge Display Technologies Ltd

Tracking what companies are doing throughout the industry allows one to identify which are leaders, and which are followers. It also identifies their level of development and commitment. This is key because it identifies which organizations should receive the majority of the resources from the startup.

The data collected thus far is used to understand the trends for the display market. Knowledge of the market size, technology trends, competing technologies, and competitors' intentions has been obtained. The next set of data to gather is on the value chain. It is important to understand the value of the technology. This is performed by, first understanding the current and competing technologies and their manufacturing processes. How does everyone do it today? Then determine what differentiates the emerging technology. Does this differentiation contain value?

Value Chain Analysis

After determining what markets and applications to go after, the next set of analysis involves the value chain. It is important to understand the value of the technology. This is performed by, first understanding the current and competing technologies and their manufacturing processes. How does everyone do it today? Then determine what differentiates the emerging technology. Does this differentiation contain value?

Excellent and Stable Visibility Performance	**Low Power Consumption**
• High color purity • Large viewing angle • Short response time (Video) • High daylight contrast • Long material lifetime (> 5,000 h)	• High material efficiency • Low driving voltage • Power efficient driving IC's • Clever driving modes
Compact Display Design	**Competitive Costs**
• Minimum module thickness • Minimum border width • Compact interconnect concept	• Simple cell process (high yields) • Cost effective production equipment • Low materials costs • Cost competitive driver IC's • Cost competitive module concept
Light Emitting Polymer Requirements	
Cell Output (peak white) 200 cd/m^2 Pixel Aperture Ratio 40 % Ratio R:G:B 20:60:20 Ave. Lumin R,B 40 cd/m^2 x 3/0.4=300 cd/m^2 Ave. Lumin G 120 cd/m^2 x 3/0.4=900 cd/m^2 Peak Lumin (M80, 128x160) R,B 30000 cd/m^2 Peak Lumin (M80, 128x160) G 90000 cd/m^2	Lifetime of Display > 10,000 hours Determined by: • Decrease in brightness (< 30%) • Increase in voltage (< 1 V) • Shift in white point (< 0.05)

EXHIBIT 11.9 Customer Expectations for an OLED Display

Source: Philips Mobile Display Systems

In the words of Jack Welch, "Keep the customer as the main focus." What are the customer's expectations for OLED technology? What are the market requirements? During October of 2002, Philips Mobile Display Systems presented in Yokohama Japan, 'Customer Expectations for OLED displays' as shown in Exhibit 11.9. Customers expected low power consumption with high performance at a low price. This provided design targets and milestones for the Poly-OLED development group.

Value Chain Analysis – *Production Value*

Evaluating the competing technologies requires determining their advantages and disadvantages. Comprehensive knowledge of the field of competition is required to best position the Poly-OLED technology. The field of competition is not constrained to the performance of a final display assembly. It includes ALL manufacturing aspects of a display.

How is it done today? A display is manufactured on a production line described by its generation. A commonly used display size for mobile phones is 2.1 inches (the diagonal distance of the viewable area). A Gen 2 production line, which has a substrate size of 400mm x 500mm, can manufacture approximately seventy-seven displays on one sheet of glass. The TFT structure is created on the glass and is patterned to optimize the number of displays manufactured. By increasing the size of the glass, the number of displays manufactured can be increased. The process cost remains relatively constant while reducing the cost of each display manufactured. A Gen 8 production line can manufacture displays on a sheet of glass approximately 2.1 meters by 2.2 meters. Exhibit 11.10 shows the assembly of components required to manufacture a full color TFT-LCD (Thin Film Transistor – Liquid Crystal Display).

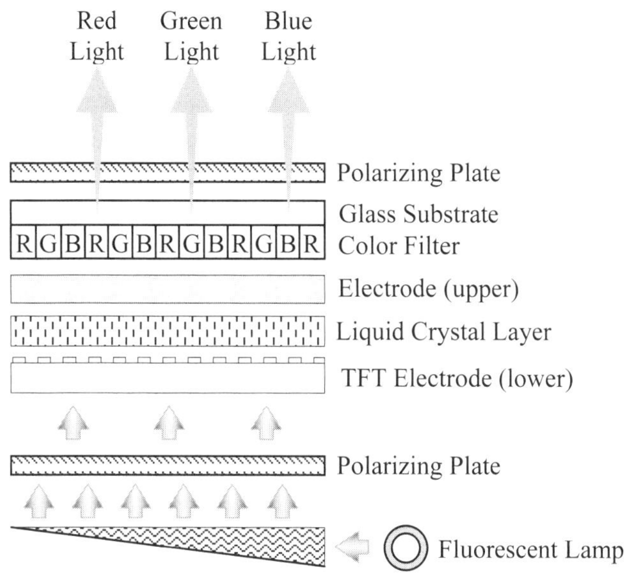

EXHIBIT 11.10 Full Color TFT-LCD Component Assembly

The number of components includes a light source (lamp) which is reflected through a polarizing plate and the remaining layers. The transistor (TFT) switches the liquid crystal allowing light to be projected through the color filter. Both the color filter and TFT electrode require precision manufacturing processes (CVD, PVD, etc.). These are subsequently aligned and assembled with the liquid crystal material. When the display is activated the lamp is always on, creating a constant power requirement.

Using a CVD or PVD process to create display features requires the precision alignment of screens to deposit material. Each feature created requires precise screen alignment to the previously created feature. When attempting to align large metal screens with dimensions in the microns, this can be extremely difficult to manage, especially when temperature is involved. To place this in perspective, the temperature effect upon a 500mm x 500mm stainless steel screen will change the dimension by 8.65 microns per degree Celsius (°C). When aligning the feature layers for a pixel size of 75 microns (300 pixels per inch), a 2°C temperature change (from 22°C (71.6°F) to 24°C (75.2°F)) will increase the length by 17.3 microns, resulting in a 23% alignment error for the pixel feature. Thus a process requiring screens, which match the size of the substrate, quickly reaches a size limitation due to alignment issues. An alternative process to apply material is spin coating. Spin coating is performed by spinning a thin flat substrate while depositing material on it, resulting in a thin uniform layer for etching. This process has obvious substrate size limitations and also requires the alignment of screens. Exhibit

11.11 provides a graphic representation of how ink jet printing can be used to manufacture Poly-OLED displays.

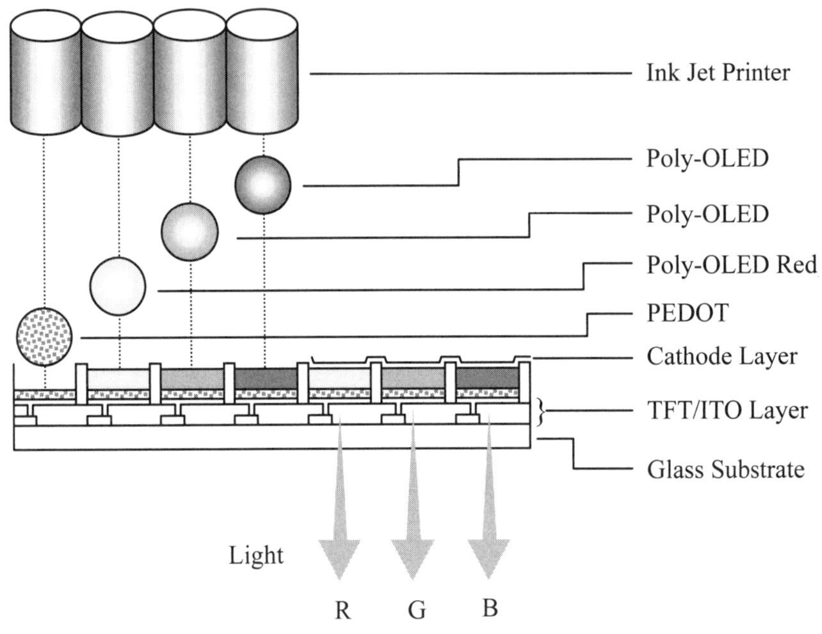

EXHIBIT 11.11 Ink Jet Printing of Poly-OLED Display

As demonstrated by Exhibit 11.11, there are fewer components to a Poly-OLED display than an LCD display. This process is additive, meaning that each layer is built up from a single substrate. When the display is activated the light required is turned on by the transistor under the pixel (emissive). This reduces power consumption using light only when needed and not all the time like an LCD display. Up to 50% of the display cost can be reduced, resulting in a display which is thinner, lighter in weight, has a wider viewing angle, and has lower power consumption. Poly-OLED also has the potential to be utilized in the emerging flexible display technologies.

Value Chain Analysis – ***Technical Value***

Developing and managing the supply and distribution channels is a required part of the process of managing emerging technologies. Each currently adopted technology, competing technology and other emerging technologies are to be evaluated. What are they? How did they originate? And what threats do they present?

On the material supply side, CDT had collaborated with Uniax and managed to sign Poly-OLED development agreements with DOW Chemical and Covion. Traditional chemical suppliers like to sell large volumes of material, train cars full. When looking forward into the Poly-OLED market, ink jet printing was changing the landscape from a 96% material waste to a very efficient 96% usage model. This meant a significant reduction in material consumption. Polyink's business model required chemical manufacturers to sell through Polyink with a value added increase in cost to the display manufacturers. The chemical manufacturers wanted to sell directly to the display manufactures and avoid sharing profit with Polyink.

Also in 2000, DuPont Displays acquired the Poly-OLED material manufacturer Uniax. DuPont contained this technology within the organization for use in their display business. This left DOW and Covion as the remaining Poly-OLED material suppliers to the display manufacturers.

There were other emerging technologies under evaluation by the display manufacturers. These are presented in Exhibit 11.12. The materials were either solution or vacuum deposition processed. They were also phosphorous or fluorescent. Other technologies such as Quantum Dots were beginning to emerge from the laboratory as well. With all the industry hype and resources spent on R&D, the major limiting factor to adoption was the lifetime of blue. Continuing the process of controlling the Poly-OLED technology, CDT acquired Oxford based Opsys Limited in 2002. Giving CDT control of all solution processed OLED chemicals. The competing materials SMOLED and PHOLED require a vacuum deposition process which has manufacturing size limitations.

SMOLED (Small Molecule Organic Light Emitting Diode) - In the late 1970s, an Eastman Kodak Company scientist Dr. Ching Tang discovered that sending an electrical current through a carbon compound caused these materials to glow. In 1987, Dr. Tang and Steven Van Slyke reported the discovery of SMOLED materials, which became the foundation for SMOLED displays produced today. In January of 2001 Kodak partnered with Sanyo to produce SMOLED displays. May of 2002 Dr. Ching Tang was named a Fellow of the Society for Information Display (SID) for his pioneering work in the discovery of organic light emitting diode (OLED) technology.

PHOLED – From 1994 to 2003, the Air Force Research Laboratory (AFRL) managed a $10 million research effort funded by the Defense Advanced Research Projects Agency. Those research efforts led to the invention of PHOLED material by Princeton University and the University of Southern California and to its refinement to batch purity suitable for display manufacturing by Universal Display Corporation (UDC) and PPG Industries. Founded in 1994, UDC has nurtured relationships with Princeton University and the University of Southern California, to conduct cutting-edge research of OLEDs and other organic electronic technologies. UDC has the exclusive worldwide licensing rights to all technology developed through these efforts. In 2000, UDC acquired Motorola's OLED patent portfolio. December of 2002, UDC announced the joint development collaboration with DuPont to develop a phosphorous-based OLED (PHOLED) material. Combining DuPont's solution based processing expertise with UCD's PHOLED technology they expected to develop a material, which could be ink jet printed. DuPont intended to use the material in their display production while UDC would leverage the licensing aspects of the material to other display manufacturers.

Solution Process	Opsys Material (Iridium material promising 30-40 L/W)	Poly-OLED (CDT Material)
Vacuum Deposition Process	PHOLED (UDC Material Manufactured by PPG)	SMOLED (Sanyo-Kodak Material)
	Phosphorous	Fluorescent

EXHIBIT 11.12 Emerging OLED Material Technologies

Pricing of the material was still an issue. The chemical manufactures were not disclosing a price for the material when purchased for mass production. They needed to understand the display industry and the value their material brought to the market.

Since material manufacturers had little or no knowledge of the cost to manufacture a display, they were experimenting with different business models. What they did know was that for an LCD, 75% of the manufacturing cost is materials. Out of that 75%, 65% was comprised of the backlight, polarizer, and color filter, none of which was required for a Poly-OLED display. They ran the concept of selling material based upon a percentage of the selling price of the display past the display manufactures. The advantage is that as the market price reduced over time, so did the material cost. If they priced the material too high, it would kill the emerging technology.

Display manufacturers had to project a possible material cost in their production planning process. Cost estimates could be obtained by using comments made by the material manufacturers. For example:

> DOW claimed they had 30 employees working on the development of Poly-OLED material. At US$200K per head, this amounted to a yearly development cost of US$6M. If DOW invested for an equivalent 3 full years prior to commercialization, this would amount to about US$20-25M including cost of capital, but not including any equipment costs, which must be amortized over 5 years (assume yearly amortization cost of $4M). This gives an estimated actual and amortized annual R&D cost of about $10M. Assuming DOW has a 50% gross margin on sales then DOW would need to generate about $20-25M/year to cover all ongoing R&D costs plus generate

a reasonable operating margin to cover Selling General and Administrative (SG&A), amortized costs and operating profit.

Poly-OLED consumption estimates for the total industry demand was about 200kg in 2005. If DOW had a 40% share then they would sell 80Kg annually. To achieve target revenues, DOW would need to charge about $250-$320/gm of material. From the above analysis, the variable cost breakeven point is actually closer to $150/gm at a DOW volume of 80 kg and production costs at 30-40%.

Integrating this projected cost to the material consumption calculation per sheet for a 3 color display using 3g (1% solution) × 3 color/substrate is approximately 0.1 grams per substrate. At US$300/gram the Poly-OLED material cost for a substrate would be US$30. This calculation is for a Generation 2 substrate, which can hold 77 to 80 mobile phone displays. This equates to US$0.38 in Poly-OLED material cost per 3-color display.

The primary competing technology for Poly-OLED is LCD. The market price for LCD displays is a moving target as it is constantly decreasing. By the time Poly-OLED material specifications are achieved, the LCD cost will be lower than it is in 2002. Covion openly disclosed that they had 64 employees in 2002 with total operating costs of US$12.5 Million and an estimated US$15.0 million for 2003. Revenues for 2002 were US$9.0 million. Revenues were from funded R&D projects and sales of material (Covion Super Yellow).

In addition to floating various business models for material, the material suppliers were also implementing blocking strategies against Polyink by claiming that ink jet printing is nothing more than a simple material deposition process, not unlike spin coating. They were encouraging display manufacturers to develop their own ink jet technology and avoid participating in the Polyink business model. As the Polyink group met with display manufacturers, they were frequently presented with this simplified perspective.

Poly-OLED's potential was the ability to manufacture larger displays with solution processing and significant lower material cost than LCD. This would achieve the thinner, lower power and higher performance pressures from the market.

Value Chain Analysis – *Production Value*

Referring to Exhibit 11.13 an Active Matrix Poly-OLED production process starts with Poly Silicone patterned glass (high temperature or low temperature). High Temperature Poly Silicone (HTPS) could be obtained from several manufacturers such as Toshiba, Hitachi and Sanyo. Developments on amorphous silicone were advancing and beginning to achieve the required switching speeds. After surface treatment, the anode layer is ink jet printed and cured. Next the substrate moves onto the deposition of the Poly-OLED with material curing steps after each color. After the substrate completes the process through encapsulation it is then cracked and broke into single displays.

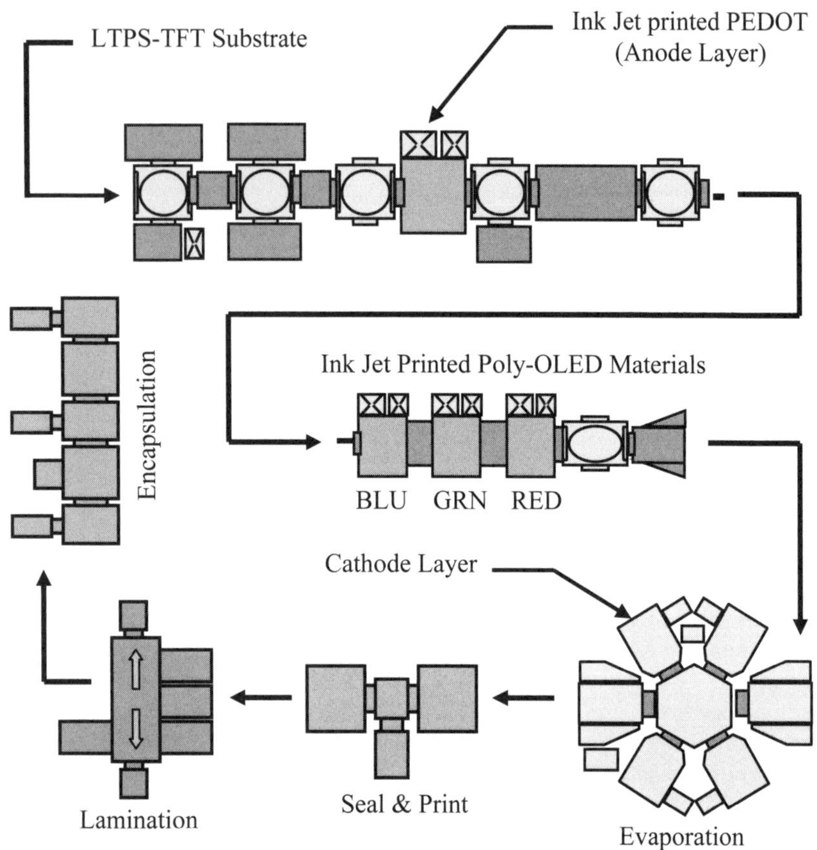

EXHIBIT 11.13 Production Ink Jet Printing Process for Poly-OLED Displays

Courtesy: Cambridge Display Technologies, Ltd.

There are far more issues to manufacturing Poly-OLED displays than depositing the material. To obtain increased lifetimes some display manufacturers were using two different cathode layer materials, which required repeating the vapor deposition process. LiF/Ca/Al worked best as a cathode for Blue, while it deteriorated Red and Green. DOW developed a Red that was compatible with LiF/Ca/Al and expected to have a Green available soon. Encapsulation continued to be an issue, and the electron transport layer between the OLED material and the TFT layer was not stable (Exhibit 11.10). The commonly used material for the electron transport layer was PEDOT supplied by H.C. Starck Inc. (USA).

Method	Main Display Manufacturers	Merits	Limitations/Faults
Mask evaporation, Small Molecule	Kodak, Pioneer, NEC, Sanyo, Sony	Able to make sample panel for mass production, 3 color	Location accuracy in large size, cleaning of toxic materials
Spin Coating Poly-OLED + Mask Evaporation SMOLED	Mitsubishi Chemical, Epson	Lower driving voltage	Limited substrate size
Ink Jet Printing (IJP) Poly-OLED	Epson, Philips	High precision and large size, multi-panels per batch, non-toxic materials	Hard to develop blue ink, luminous uniformity, lifetime, requires exclusive IJ head design
IJP RED GRN of Poly-OLED + Solid Print evaporation of SMOLED BLU	Epson, Sumitomo, LG Chemical	Able to use Blue ink of SMOLED, Higher light emission efficiency	Hybrid device structure need IP right from Kodak
Laser Induced Thermal Imaging (LITI), SMOLED	Samsung, 3M, DuPont, Kodak, Epson	High precision & large size, higher light emission efficiency	Special property, control of transcription face
Poly-OLED shadow mask spray	ULVAC	Simple Process	High precision Limited substrate size
IJP of SMOLED	Kodak	Without metal mask	Special property
Poly-OLED, Gravure offset printing	Toppan	Ability to make large size	Low precision

EXHIBIT 11.14 Summary of Poly-OLED Material Deposition Processes

Courtesy: Cambridge Display Technologies, Ltd.

Display manufacturers quickly learned that beyond the ink chemistries, depositing the Poly-OLED ink was only an integral part of the over all solution and the knowledge required to perform the secondary operations such as drying and encapsulation were equally important. In exhibit 11.13 deposition of the Poly-OLED material is performed by ink jet printing. The process requires very precise deposition of Poly-OLED material onto substrates. Controlling the drop size, placement, and volume into a micron sized pixel, has to be balanced to achieve the desired post processing effect. The ink jet printing process can precisely deposit Poly-OLED material onto substrates with very high material use efficiency.

In an attempt to reduce product costs, display manufacturers increase substrate sizes to Gen 4, Gen 5, and up to Gen 10 size substrates. Spin coating or laser transfer is extremely difficult with substrates of this size. Exhibit 11.14 compares different deposition processes.

Most industry adopted manufacturing methods for Poly-OLED are either limited in substrate size or required hazardous material handling. Ink jet printing did not have either of these issues. Ink jet printing had other issues such as secondary processing to obtain uniformity and custom head design and controls. Ink jet printing reduced the number of processing steps resulting in smaller facility requirements and lower capital equipment costs.

Financial

CDT's funding started with very limited seed venture capital in 1992. Their early business model was to manufacture Poly-OLED displays and compete in the display market. The display market was asking for a three-color solution to manufacture displays and the material needed further research to achieve the required specifications. In 1997 CDT obtained another US$10 Million from a financial group headed by Lord Young of Grafham, former UK Secretary of State for Trade and Industry. Later in the same year, Intel, the semiconductor manufacturer, also invested in CDT. In 1999 Kelso Investment Associates and Hillman Capital acquired a majority interest in CDT for a total of US$133M. Kelso and Hillman are both private equity funds, based in New York, USA. Additionally they provided additional funding of US$16M directly to the company to finance ongoing research and development activities.

While having dinner with his wife in Monterey CA, attending a Stanford Resources (currently iSuppli) conference in November of 2000, David Fyfe over heard a conversation at a neighboring dinner table. A couple of gentlemen behind him were discussing CDT. The first gentleman asked the other if he knew of CDT? The second gentleman replied, "Yes", he did. In fact he had recently called on the company in Cambridge. "What did you think of them?" asked the first gentleman. The second gentleman replied, "A bunch of very bright scientists, but I don't think the company will go anywhere. They don't have any practical experience." This is a clear message to all managers who undertake the position of managing emerging technologies. Stop the shoestring budget hobby efforts emulated by university laboratories and turn it into a real company. Build the required infrastructure to solve technical problems, and the reputation required to build collaborations. Enlightened with this comment, David set out to obtain additional funding to move CDT beyond a laboratory-based company and into a material and process development company. In 2001, a further US$28m was raised from shareholders. These funds were to finance construction of the Technology Development Centre in Godmanchester UK, and substantially increase the burn rate. On April 10th, 2002, CDT, with Lord Sainsbury, the Minister for Science and Technology in the UK, officially opened the 1,750 square meters, Generation 2, Technology Development Centre in Godmanchester, U.K. Construction of the advanced lab came in on budget and on time at a cost of US$25 million.

With the increase in funding CDT grew from the reputation of being a small bunch of researchers into an organization with proper resources.

Risk Identification

Risks are constantly evolving. Some can be expected and some appear without warning. Items such as lifetimes, efficiency and luminance are specifications which are defined by the market and can be mitigated. Other risks such as supply chain dynamics can occur without warning and one must be prepared to act. In the case with ink jet printing of Poly-OLED, CDT witnessed a pending loss of market control to one of the material suppliers. CDT stepped in and acquired Litrex, further enabling the expansion of Poly-OLED development.

Exhibit 11.15 identifies risks associated with the Poly-OLED technology up through the year 2002. The primary risk drivers for Poly-OLED are material maturity, channel stability, unstable market dynamics and blocking strategies from collaborators and channel members.

Maturity of the technology limited adoption. Encapsulation and the lifetime requirements were proving to be the significant milestones difficult to achieve. Additional research was required from both CDT and display manufacturers.

Portfolio Position

Poly-OLED can be used within numerous applications. CDT has retained focus on display market applications not allowing themselves to be distracted or their resources diluted working on non strategic applications. By analyzing the market, strategic collaborations were nurtured with those who are best positioned to adopt the technology. These collaborators possess channels, critical display manufacturing technology, and resources for R&D. They track the market. They are aware of technology trends and profit reductions of maturing display technologies. For a display manufacturer Poly-OLED can be in alignment with their technology road map adding the opportunity to create the next level on their technology S-curve. Display manufacturers will determine which sector Poly-OLED technology will provide sufficient differentiation to achieve rapid adoption.

Risk Dimension	Risk	Rating	Mitigation
Government	Regulatory – Chemical hazard regulations	Medium	Monitor
	Cultural – Creating National projects removing foreign involvement	Low	Position with key organizations
	Business Design – IP filing & enforcement	Low	File IP and monitor
	Liability – Product	Low	contracted
Market	Competing Technologies	Medium	Monitor & Acquire
	Collaboration Development	Medium	promote value proposition
	Business Design – IP licensing, requires high burn rate	Medium	Push for design wins
	Cultural – Multinational methods of doing business	Medium	Local & International Talent
	Blocking Strategies – From customers and suppliers	High	Constant management
Organizational	Distribution Channel	High	Constant management
	Dependence Upon External Resources	Low	Organic R&D
	Speed of Change	Low	Small organization
	Available Capital	Medium	Investor involvement
	Management Structure/Resources	Low	Recruit talent from market
Technology	Maturity – Requires additional R&D Core IP expires in 2009	High	Investment & Collaborations
	Alignment with Core Competence	Low	Review & focus
	Manufacturability	Low	Monitor
	Projected Cost	Low	Monitor
	Supply Chain (Maturity/stability)	High	Constant Management
	Competing Emerging Technologies	Medium	Monitor

EXHIBIT 11.15 Poly-OLED Phase 2 risk identification table

IP Position

By the end of 2002, CDT was in control of 131 patents around their Poly-OLED technology with several applications pending. When UDC acquired the OLED patents from Motorola they obtained more than 70 patents and held many more around their PHOLED technology.

One of the critical patents being leveraged by the material manufacturers was the patent filed by Princeton University (US6087196) for "Fabrication of Organic Semiconductor Devices using Ink Jet Printing". This patent was assigned to UDC. UDC subsequently applied for and was granted a world patent (WO 99/39373). Material manufacturers raised this issue to CDT, Epson and the display manufacturers. This patent had the potential of negating CDT and Epson's claim on ink jet printing of OLED displays, which is fundamental to CDT's strategy. Epson's response to the UDC patent was that it actually covers a wide claim of "making OLED using IJ" which was applied on January 30, 1998. However, Epson had previously applied for the following three similar patents.

1. Making OLED devices using IJ (basic) Applied on Nov 25, 1996.
2. Making OLED devices using IJ process and also using bank, applied on Sept 19, 1996.
3. Making TFT from liquid phase (with some description on IJ process)

It was understood that because these three applications pre-dated UDC's, the UDC patent was not valid and an interference process was submitted to the US Patent Office. Subsequently a material specific ink jet printing patent was issued to UDC in January of 2006 (US Patent No. 6,982,179, titled "Structure and Method of Fabricating Organic Devices"). This complicated issue was resolved with a settlement and license agreement between Epson and UDC executed in July of 2006.

CDT has managed to develop and control a significant number of patents. These patents included core material patents, processing patents and application patents. Their comprehensive list includes world patents and places them in a strong market position.

Business Design

CDT's original business design was to compete in the display market by manufacturing displays in the UK, building the channels to market and capturing the maximum value for each device. Realizing the extensive capital investment required competing in the display market and not possessing display device technology; CDT converted their business model to an IP licensing model. CDT repositioned into a technology enabling and solution provider, collecting fees from licensing and research projects. Exhibit 11.16 outlines CDT's key dimensions with each section of their business strategy. Focus had to be on channel dynamics and IP management.

Section	Key Dimensions
Customer Selection	• Display manufacturers • Material manufacturers • Process development organizations
Value Capture	• Licensing • R&D contracting • Growing IP portfolio
Differentiation/ Strategic Control	• Price (solution processing, fewer components) • Technology (emissive & organic) • Intellectual Property • Thin, Low power • Can be used in flexible displays • Customer Relationships
Scope	• Display & lighting products • Material, processing and device solutions • International presence • Industry wide collaborations
Culture	• Education through collaborations & conferences • Local talent to manage language barriers • RoHS* compatible, 3rd party involvement with government projects (Future Vision, etc.)

* The Restriction of Hazardous Substances Directive (RoHS) 2002/95/EC

EXHIBIT 11.16 Poly-OLED Phase 2 Business Design Dimensions

Poly-OLED Phase 2 Summary Statement

By the end of 2002, CDT had acquired the technology from Opsys and they had three primary material suppliers performing research and developing materials. With the acquisitions of Litrex and the cross license agreement with Epson they had secured control of the preferred method of manufacturing, ink jet printing. They had ongoing collaborations with most of the display manufacturers and Philips launched the first product containing Poly-OLED technology in their electronic shaver. With the investment in the Godmanchester facility, CDT expanded their capabilities to provide technology demonstrators and increased their processing IP portfolio. The Godmanchester production line included ink jet printing systems from Litrex Corporation.

There were an increasing number of announcements demonstrating successful technical achievements for Poly-OLED. CDT increased their patent portfolio and focused their business model. This market acceptance also provided the opportunity for CDT to obtain further investment.

All parameters of the phase 2 analysis have been addressed. It demonstrates that an extensive amount of market and technology knowledge is critical to market adoption. Without managing trends and threats one can easily loose market control resulting in lost opportunity and lost value to the organization.

Phase 3 – Quick Test Market

Phase 3 involves the implementation of a business strategy derived from market and technology data. There is not necessarily a clear cut starting point when phase two transitions to phase three. Rather, there is more likely to be an overlap between the phases.

Phase 3 – Quick Test Market		
Parameter	**Comments**	**Report**
Collaborators	Summary Comment by Project Team	Document
NDA, JDA, JVA	Summary Comment by Project Team	Document
Intellectual Property	Summary Comment by Project Team	Document
Risk Management	Summary Comment by Project Team	Document
Portfolio Position	Summary Comment by Project Team	Document
Opportunities & Requirements	Summary Comment by Project Team	Document
Analyze Results	Summary Comment by Project Team	Document
Business Design Review	Summary Comment by Project Team	Document
Summary	Management Review & Summary	Document

EXHIBIT 11.17 Phase 3 Quick Test Market Review Table

Exhibit 11.17 is the review table for phase 3. This table monitors the continuing growth of CDT and test markets the business strategy.

As the number of collaborators continued to increase, regular review of contract agreements is a good practice. NDAs were reviewed for expiration dates. JDAs were reviewed to ensure the scope still captured the evolved developments. The number of patents filed continued to increase as did license agreements. Market acceptance was a positive sign reflecting confidence in their risk mitigation and the business model.

In Phase 4, the team was expanded to encompass the greater skill sets and investment required to bring the technology to market.

By Phase 5, the action plans for production development should have been completed, which brings the emerging technology to a lower risk platform and increases the probability of returns for investors.

CDT's quick market test and action plans implementation of phase three to phase five took place up to year 2002.

By the end of 2002 CDT had an extensive staff with a clear direction defined. The business strategy was being rolled out and the company had evolved into an efficient development engine, coordinating numerous exciting and promising multinational collaborations.

The remainder of this section will disclose events which occurred from the beginning of 2003 through acquisition in 2007.

2003 through 2007

As Poly-OLED entered into 2003, interest dwindled amongst the display manufacturers and investors. Event though CDT announced a lifetime breakthrough of 20,000 hours for blue, display manufacturers were having difficulty reproducing the numbers. They were also observing the continuous reduction of market prices for LCD displays. This reduction in LCD display costs made them begin to question the future profitability for Poly-OLED. Philips continued manufacturing their razor with the Poly-OLED display and demonstrated an OLED display which doubled as a mirror on their 639 mobile phone. In the meantime, CDT's Godmanchester facility achieved full operational status and CDT began accepting orders for 1.7" displays. Interest in Poly-OLED had slowed but the market still moved forward.

Kodak's material also started to enter the market in early 2003 as they introduced the first digital camera with a SM-OLED display, a competitive emerging technology. Kodak continued to obtain license agreements with display manufactures. This increased market pressure on Poly-OLED.

ULVAC Inc. (Japan) is a large chemical vapor deposition equipment supplier to the semiconductor industry. Their equipment is used to manufacture SM-OLED displays. Monitoring the emerging market for Poly-OLED, ULVAC determined they needed solution processing capabilities. They experimented with developing their own and explored the possibilities of collaboration with several companies. Litrex was the only viable ink jet printing solution available on the market and was rapidly gaining market share. In August of 2003, CDT sold 50% it's interest in Litrex to ULVAC. ULVAC would acquire the remaining 50% in Q4 of 2005.

By May of 2004 Sony announced the development of a 24.2 inch SM-OLED panel and Epson demonstrated a 40 inch Poly-OLED panel. Epson announced commercialization plans of up to 41 inch Poly-OLED televisions by 2007. A few months later Sony announced the development of an 11 inch SM-OLED television. The market struggle between SM-OLED and Poly-OLED continued with both technologies receiving technology awards and obtaining license agreements. On December 7th of 2004 Epson reaffirmed their production plans for Poly-OLED televisions by 2007. Nine days later CDT went public, offering 2.5 million shares for US$12 per share. The market closed the first day of trading at US$11.27/share.

2005 proved to be another dynamic market shift for Poly-OLED. On January 28th Merck KGaA sold their electronic materials division to BASF for €270 million and 11 days later announced the acquisition of the Poly-OLED division from Avecia for €50 Million in cash. This acquisition included their Covion subsidiary. DOW Chemical Company decided to halt their investments in Poly-OLED and placed Lumation™ up for sale. Lumation™ was acquired by Sumitomo Chemical on May 16th. This reduced the number of Poly-OLED raw material suppliers to two. Merck KGaA was just beginning to gain experience with the technology while Sumitomo had several years behind them. From an IP perspective, Sumitomo held patents and managed to obtain critical ones with the acquisition. CDT noticed this imbalance of IP and eight days after Sumitomo's acquisition of Lumation™, CDT and Sumitomo announced their joint venture named Sumation™. This consolidation in material supply was followed by Philips' sale of their Poly-OLED division to the OTB Group in August of 2005. The OTB Group created the OTB Displays division with a business model of selling complete Poly-OLED production solutions, enabling companies to manufacture Poly-OLED displays. One could manufacture Poly-OLED displays with a license from CDT, material from Sumation™ and a production line from OTB Displays which included ink jet machines from Litrex. This structure ensured CDT had the ability to promote display production, without IP issues, and placed the Poly-OLED technology on the definitive path in the display market.

Yet with the material suppliers in flux and the sale of the Philips production line, investors began to down grade CDT's stock. During 2005 it managed to fall below US$6.00 (a substantial 53% lower than its IPO price)

The joint venture Sumation™ proved to be beneficial as the combined technologies resulted in achieving record lifetimes for Poly-OLED (>500K hours for red and >150K hours for blue). In conjunction with the announcement of achieving record lifetimes, CDT ended the year with an additional funding of US$17.5 million from a private stock placement with institutional and other accredited investors.

2006 continued with improved material properties. CDT also continued to receive additional technical awards. Processing improvements were enhanced with higher resolution ink jet machines from Litrex.

In April 2007 at the Flat Panel Display (FPD) show in Japan, Toshiba Matsushita Display Technology (TMD) demonstrated a 21 inch Poly-OLED television. This exhibition of TMD's Poly-OLED display capabilities marked a significant milestone in the advancement toward commercial adoption of Poly-OLEDs as the next-generation display technology and demonstrates one of the key differentiating advantages of Poly-OLED technology over SM-OLED, which has long been considered to be the more scalable and practical OLED-based solution for larger substrate sizes.

In May of 2007, MicroEmissive Displays (MED) announced the completion of their production line in Dresden and that they will begin their initial production in July. MED reported back orders of US$10.5 million.

On July 31st, 2007, Sumitomo Chemical announced the acquisition of CDT for US$286 Million.

CHAPTER SUMMARY

An emerging technology needs more than a list of potential applications, it also requires proof of concept for each application. Creating proof of concept can be either a fundamental or complex approach. In the case of displays, a fundamental proof of concept is creating a simple 3x5 array, while a complex approach includes manufacturing a complete functional display. The complex approach enables enhanced IP opportunity in areas such as processing and wave-form generation. Substantial investment is required to create such facilities and develop associated IP. CDT made the critical decision to build their production/development facility, which resulted in additional IP and a better relationship with collaborators. It enabled them to demonstrate fully functional devices, placing the technology in the path of the market.

Ownership of intellectual property results in market control. CDT built a comprehensive IP portfolio and continued to grow it over time. With an IP business model they were *required* to build IP. This requires a high burn rate with a substantial R&D budget.

An in-depth market analysis was not originally performed by CDT. Identification of market leaders and strategic collaborators occurred as a result of continual networking and multiple presentations at technical conferences. CDT received numerous inquiries as they continued to present achieved milestones, and subsequently established collaborations with display manufacturers. Their alliance strategy evolved along the way with a targeted approach to select collaborators that offered the highest probable route to success. Advancements and market acceptance greatly improved when CDT began to develop close collaborations with market leaders.

This case study was written to demonstrate the in-depth market knowledge beyond the technology that is required to manage an emerging technology. Not its function, or market value, but also its path to market and the players involved. Managing an emerging technology requires the following:

1. Highly skilled management team.
2. Relationship building, relationships with investors and collaborators
3. Continuous monitoring of the market for threats from other emerging technologies and channel members positioning for control
4. Knowledge of competing technologies and their processes
5. Substantial investment for realistic development and the ability to adjust to changing market conditions (potential acquisitions)

This is the front line position for the organization, both offense and defense. Decisions here will make or break an organization. The individual(s) performing this task will also require support from a team of highly qualified technical experts in the organization.

Another critical aspect revealed by this case study is that numerous market threads come together as an emerging technology moves from the beaker to market adoption. Poly-OLED originated in a laboratory where it required interest from investors and the market before it could be driven to adoption. Manufacturing processes developed in Rastor Graphics were acquired by Gretag, subsequently acquired by CDT and later sold to ULVAC. Material manufacturing collaborations such as DOW Chemical Company and Covion were cultivated and subsequently acquired. Other emerging materials evolved such as Opsys and were acquired. These are just a few examples of the market threads which came together during the evolution of Poly-OLED. Each of these threads played a significant role in both market dynamics and IP management. Without managing both the IP and processing technologies throughout the supply chain, CDT may have allowed another organization to obtain control and a greater portion of revenue from the value chain. CDT successfully managed these threads resulting in control of an extensive number of patents and enabling technologies. With the renewed market attention from the announcement of Sony's 11 inch television production plans, other organizations accelerated their production plans. This resulted in a market pull for the technology and strategic positioning for Sumitomo through the acquisition of CDT.

Chapter 12

Summary (5 take-aways)

Success is more a function of consistent common sense than it is of genius. - An Wang

Too often the shortcut, the line of least resistance, is responsible for evanescent and unsatisfactory success. - Louis Binstock

Adoption of emerging technologies is frequently met with failure or do not achieve expectations. Often the technology can be disruptive or it can enable a completely unrelated or unforeseen opportunity. When the technology becomes mature enough to disclose to selected collaborators, the final application may end up not being what was originally planned. Identifying the required collaborative technologies and the organizations that own them is crucial in the process of moving the technology from the laboratory to the market. Only by understanding the fundamental technology and its role within the technology value chain, can an effective business strategy be architected. Deciding whether to pursue the venture or not depends upon numerous factors all coming together.

This book discussed the management of numerous critical factors which contribute to the successful adoption of an emerging technology. Its goal is to demonstrate that once the Eureka has occurred, there still remains an extensive amount of work to achieve success. Gartner discussed how an emerging technology undergoes a hype cycle and the commitment required after the trough of disillusionment. Pauchard demonstrated with his Innovation Engine that most innovation is not planned like a new product with a predefined specification. But rather it *occurs* when technologies and applications cross or merge, exampled by the emergence of 3M's Post-It® note pad.

Issues around technology transfers and how to manage the development of technologies were discussed. Developing clearly defined project plans and valuing the management of the proper stages of development.

While the technology is under development, strategic marketing processes were provided designed to identify potential target market segments. This section provided the tools to evaluate markets and business strategies in detail. When completed, the management team will thoroughly understand where to best position the technology, how it is differentiated and the risks associated. The chapter finished with business model profiling that could be communicated to stakeholders and throughout the organization.

It is business development which places the technology in the market. This business function aligns technology development and the organization to achieve market adoption, via the provided customer discovery process and gap analysis. The often overlooked significance of cultural impacts was exemplified for aspects of industry, regional, communication, macro effects and product requirements to enhance cultural adoption. Connectors located throughout an industry can be leveraged to help accelerate adoption rates. The chapter on business development provided processes to develop the elevator pitch and identified the significance of developing a communications strategy to stakeholders and the market.

Communicating the emerging technology often requires the execution of contracts such as NDA's and JDA's. Chapter 6 provided high level insight into the issues that need to be managed. It discussed intellectual property from a business perspective coupled with management for strategy. The goal of the chapter on legal perspective was to clearly identify that there is no substitute for hiring experienced sound legal council.

It is crucial for innovators to understand investors and effectively manage the relationship. Innovators, being the creator of the technology, often want to retain a controlling interest in the startup company. They also expect their exit share of ownership to be equal to their entrance share. However, these approaches can inhibit progress with investors. Innovators need to realize that as projects progress; financial returns vary according to the remaining level of risk. The financial chapter explained the process of the different funding cycles and associated dilution from a venture capitalist approach. It outlined how to mange cash flow and align it the organization. The financial chapter finished with Do's and Don'ts from the investor community.

In a startup organization it is the CEO who is held accountable for delivery of numbers and milestones. This is best achieved with effective organizational management. At the heart of every organization are its core values. It lays the foundation to the organization's culture. Core values reflect the corporation's intent and image to customers and to the local community. A company's mission encompasses what will be done and how to know when success has been achieved. It outlined the importance of creating a high performance team and provided processes to build and manage trust. Organizational management also provided solutions to manage FUD (Fear, Uncertainty and Doubt), which can greatly hinder the progress of any organization.

The chapter on corporate perspectives provided insight into how large organizations view emerging technologies, their effect on the organization, and future offerings. It identified processes and methods used to align research and development with the corporate strategy. Brand portfolio management and product portfolio management demonstrated how different business processes interrelate and how emerging opportunities are measured.

As organizations are frequently approached by emerging technologies, processes have evolved designed to manage them. From Eureka to adoption, Chapter 11 presented a comprehensive framework for managing an emerging technology. It provides a phased approach enabling executives to make informed decisions. This process is inclusive of check lists and evaluation tables with the intent the reader may expand upon the process, or modify to fit their requirements. This process is exampled via the in-depth Poly-OLED case study located in chapter 11.

On a positive note, exhibit 12.1 identifies with select technologies, that the historical adoption rate has a decreasing trend. This is due to numerous factors, most of which are an increased understanding and improved management of the required success factors. The telephone took 68 years to achieve a 50% adoption rate while the internet only took 9 years. Leveraging existing infrastructures, the adoption rate for new technologies can be reduced. iPOD®'s success has been attributed to a positioning the technology to leverage the global transition to an experiential economy, and extensive collaborations. iTunes® was developed through strategic collaborations to enhance the user experience. Effective marketing campaigns and combined packaging of several emerging technologies into a single product were also instrumental in the success of iPOD®.

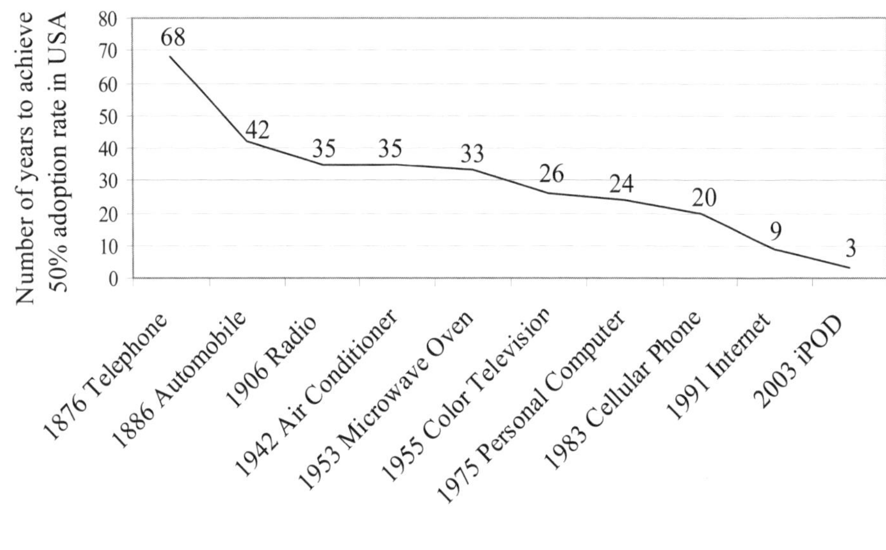

EXHIBIT 12.1 Time of adoption for new technologies

Source: Ramon Marimon, "University Research Business and Local Development", Luiss Guido Carli University, December 4, 2003

Discussed in the introduction, and reproduced in exhibit 12.2, was how many emerging technology managers focus on identifying the potential buyer who can use the technology in their process. Some focus solely on the *design win*. Simultaneously, corporations often rely exclusively on their direct suppliers to be the provider of emerging technologies. These two approaches generate a gap; where as corporations are not aware of lower level technologies that can enhance their product offerings, while emerging technologies are not aware of potential market applications beyond the product level. Bridging this gap can provide access to "killer applications" for the emerging technology and provide enhanced user experiences for service level providers.

Targeting the corporate or end user for adopting the technology has the ability to reduce the time to market adoption. For example, suppose the emerging technology reduced the energy consumption of a display by 35%. Targeting the display manufacturers sounds like a reasonable approach. However, based upon experience, one would be faced with Not Invented Here (NIH) and disbelief of the technology claims. By taking the technology to a market leader such as Nokia or Dell and convincing them of the technologies differentiation, they in turn would go to their suppliers and encourage them to adopt the technology (pull through approach).

For a corporation searching for the emerging technology which will provide them a differential advantage, relying on their supply chain may not be the best solution. This can be exampled by an emerging market application for construction. Over the past few years, the United States has experienced bridge failures. Some of which have resulted in loss of lives. As a result, the United States federal government has imposed a requirement that every bridge is to be inspected every two years. According the Federal Highway Association, there are approximately 598,000 bridges across the USA. This number excludes public transportation and foot bridges. Inspection costs can run US$50,000 per span. Suddenly the departments of transportation for every state are burdened with an unbudgeted expense.

A bridge construction company could recognize this problem and observed opportunity. For simplicity a fictitious company will be exampled (B&R Construction). Their business model is to build bridges (contract for hire). This bridge issue is in their sector and they are curious if an automated technical solution could be provided. B&R Construction consults an individual who is heavily networked throughout the emerging technology sector. They inquire to the possibility of a technical solution that can monitor a bridge's structural integrity remotely. There are bridge monitoring solutions readily available in the market. However, each requires extensive wiring of the bridges. B&R does not want the expense of wiring all the bridges. The advisor quickly outlines a collection of technologies, if pulled together, would provide a solution. Strain gages can be connected to the bridge. The data from the gages can be transmitted via a Zigbee or V-Wave chip to a node at the end of the bridge. Several bridges can be networked together by setting up a WiMax network, which would send the data to a central control room. It doesn't take much to imagine that there will be a lot of data. A company named GroovSQL™ has demonstrated the highest performance for managing real-time data. They can manage an amount of data on a single $50,000 server that any comparable solution would cost $500,000. Power to the sensors can be provided via photovoltaics. For sensors which are not exposed to light, PowerCast has a remote power delivery system. Event level artifacts could be filtered at the bridge prior to delivery to the system. The technical solution would work for either strain gages or accelerometers. However, with emerging accelerometer technologies from sources such as Jyrate or Hewlett-Packard, the real-time vibration could be coordinated to a CCD camera monitoring traffic. If a vehicle exceeding the load limit crosses the bridge, the camera can take a photo and send the driver a fine (revenue generation). In the event of a seismic occurrence, the system could immediately determine the integrity of the bridge. This solution does not contain research. It is an integration and deployment solution. A properly architected solution would have an unmatched cost/performance solution. In this example B&R found a solution by communicating across sectors and bridging channels.

The following emerging technology roadmap applies to every technology sector. It is the consumer or government agency who makes the final decision of adoption. This book identified, through a series of case studies, best practices from the market and from literature to successfully manage emerging technologies. Bridging the gap between these two different perspectives opens the door to *Killer Application* opportunities.

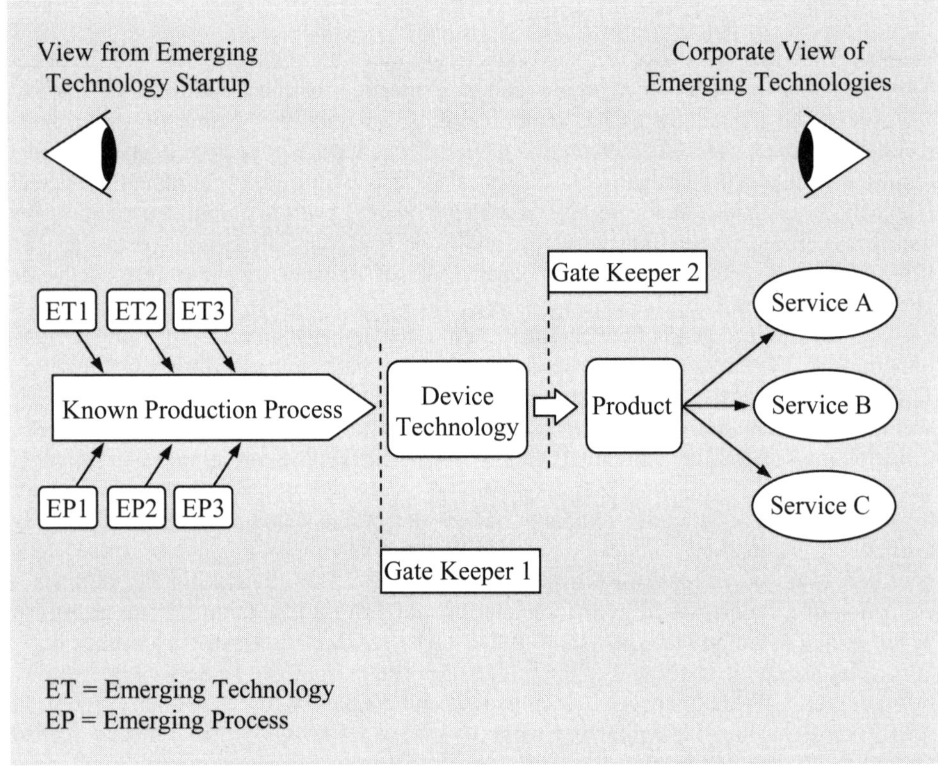

EXHIBIT 12.2 Emerging Technology Market Roadmap

The five take-aways

This section is for the reader who will pick up the book and skip to the final summary, expecting to gather comprehensive knowledge from a few pages. In as much as this text falls short, if successful management of emerging technologies could be summed up with only five take-aways, such a book would have been written long ago. History would be filled with an exhausting number of innovators easily achieving their aspirations, failures would be rare. However history has proven this to not be the case. The following is a feeble attempt at identifying five take-aways:

1) Know your value proposition to the customer and listen to the customer. Look beyond your customer and be open to alternative opportunities, which can create competitive advantage.
2) Know the market, and equally important, know the channels and who owns them. Listing a big market is not sufficient. Identify which sector(s) of that market have the highest probability of adopting your technology. Who are the leaders of that sector? How big is that sector? What is your anticipated adoption rate and how will this be achieved?
3) Know your technology and its value within the technology value chain. Know all differentiating aspects of competing, emerging, and currently adopted technical solutions. Do not underestimate adopted mainstream technology improvement possibilities, for they have the resources to perform continuous improvements.
4) Leverage, manage, and value your Knowledge Networks. Pay attention to culture!
5) Formulate a highly skilled team and empower them.

Six, is the three legged stool. Many investors have expressed frustration with innovators who claim their technology has more value than the money. Likewise, many innovators express frustrations with investors who claim their money has more value than the technology. In reality, both need each other to grow. Yet when they join together they function like a two legged stool, not very stable. They need a third leg, a jointly agreed upon growth and exit strategy with a highly skilled team who can accelerate technology and market adoption.

In closing, the author heavily emphasizes this compilation of best practices was derived from extensive research and global exposure to innovators, across numerous sectors, attempting to manage an emerging technology. The effect of driving an emerging technology to adoption is similar to putting together a puzzle. Frequently pieces of this puzzle are overlooked and not managed. The objective of this book is to add to the knowledge base of managing emerging technologies, enhancing successful adoption rates. Processes provided can be implemented and modified to achieve better alignment within your organization.

Chapter 13

Open Discussion

You need to learn how to use your ears and your mouth in the proportion God gave them to you – Robert E. Nelson

This chapter is provided as an open discussion on currently emerging technologies (at time of publication). Numerous technologies have crossed my radar; only a selected few will be mentioned due to their differentiation. It is impossible for any one individual to claim they are aware of all emerging technologies, this is why we network. Professionals who analyze emerging technologies spend a lot of time traveling and investigating technologies across numerous sectors. These positions are likened to that of going to see someone's new baby. Somewhere in the globe, a professional has spent years nurturing, incubating, taking prenatal vitamins all in anticipation of achieving their Eureka (their baby). News gets out about the new baby and villagers come to see the baby. The innovator is proud, holding the child up to the light. The innovator has dreams of the child becoming a star athlete, an educated scholar perhaps whining a Nobel Prize, or following in their parent's foot steps continuing the family business (international market adoption, disruptive technology, enabling emerging markets, etc). It is the technology analyzers job to go look at the baby and perform a feasibility analysis on it. Most often we look at the baby and say, "That's a lovely baby you have there". Sometimes when we look at the baby we realize it's a clone, a duplicated technology. It could have been derived independently, but the reality is, sometimes it's a clone (a "Me too" technology). Then there are the times when we look at the baby and we think, "Oh, that's where the phrase, *only a mother can love* comes from". But you can't tell a parent that their child is ugly, you can't tell them it's a clone. If you do, the parent will become defensive and attempt to discredit your opinion as ignorant. In those situations it is best to make a polite comment and recommend they visit another parent. Persuade them to look in the direction of the lovely children. Provide a hint for them to go look at a competing technology to help them obtain a better understanding of their true value proposition.

In this position of investigating emerging technologies and markets, we are often surprised at how frequent identical technologies are independently developed in parallel. While the innovator may believe it to be original, it can also be an exact duplicate of a technology that is in production. This is insight into why one should file their patents early. CIT (Conductive Inkjet Technologies) developed a process of inkjet printing an organic seed layer that binds to a polymer substrate. Using an electro less plating process, copper is then built up on the substrate to create a patterned conductive trace (entered my radar 2002). Developed in 2006 and announced at IDW 2007, Epson announced an inkjet printing process which patterns an inorganic seed layer then subsequently building up copper with an electro less plating process. October 2007, Averatek enters my radar; they have developed a patterning process of using inkjet printing to deposit an inorganic seed layer, then subsequently using an electro less plating process to built up copper to a desired thickness. While Epson and Averatek are making announcements, CIT has been manufacturing systems and placing them in production lines.

The baby has to grow up. It needs to learn to walk and eat by itself. It needs to learn how to read and go to school. It needs to learn how to function in society, learning the differences between right and wrong. The baby will transition from infant to child and then adult, as it grows to become a valuable contributor to society. When the parents of these new born babies decide to show them to the world, they develop a growth plan (business or marketing strategy). Often these innovators pull together a PowerPoint presentation describing their technology and associated potential market applications. Invariably, somewhere in the presentation there will be a slide; intentionally created to provide the opportunity to drop a quote from the latest popular business book (Disruptive Technology, Tipping Point, Crossing the Chasm, and many others). Many business related aspects are often overlooked, providing the inspiration behind this book. A desire to provide information to innovators that could help them achieve their goals and aspirations.

As an open discussion chapter about emerging technologies a relaxed approach will be presented, discussing selected technologies from a small set of analyzed sectors.

Materials

Most emerging technologies evolve at the material level. It is these advancements in materials that enable changes in processing and applications. As a basic rule, if the material in not RoHS compliant, commercial applications will not adopt the technology. However, Government or military applications might.

Quantum Dots (Q-Dot)

Discovered by accident, most are comprised of heavy metals. They are a Nano ring that when a lower frequency light (UV) is shined upon the Q-Dot, an electron is trapped and resonates, emitting a higher frequency light. By changing the diameter of the ring, the color of the output light can be changed.

There are display companies like QD Vision, founded on technology developed by Seth Coe-Sullivan, who manufacture displays using Q-Dots. The commercial industry is resistant to adopt Q-Dots due to the existence of heavy metal material. Yet, military applications are taking notice. Q-Dots have also been used as bio tags to identify varying types of cells, mostly cancer. By attaching a protein to the Q-Dot, which is known to bind to cancer cells, and placing it in the blood stream, the cancer cells can then be identified by shining a UV light on the tissue. However, heavy metals cannot be used in this application. Two companies have crossed my radar that can successfully make heavy metal free Q-Dots. The most advanced is Nanoco located in Manchester UK. A university spinout founded on the technology developed by Nigel Pickett and Paul O'Brian. They have developed a solution processing technology to manufacture the Q-Dots. Nanoco's technology is a much lower cost proposition than the common laser based manufacturing methods. Q-Dots can also be used in decorative applications such as placing them in coatings for devices such as mobile phones. Using a UV LED source within the phone, the exterior surface could light up the phone in a different color when it rings. There are many other novel material applications for Q-Dots.

Printed Electronics (PE)

To properly discuss this topic it needs to address the conductive inks used in the printing process, the printing process and the secondary processes used to cure or anneal the inks. Back in 1996, Epson inkjet printed the first OLED display and demonstrated the display to Cambridge Display Technologies. Also in 1996, Epson placed an inkjet machine in the Cavendish Laboratory at Cambridge University UK. There Epson and Cambridge University created the first inkjet printed transistor. That technology was spun out to what is now called Plastic Logic Ltd. These technical announcements in conjunction other milestones around the globe, have been accredited for starting the printed electronics industry. A significant amount of research has been targeted at printing TFTs on varying substrate materials. This has resulted in significant advancements in materials. The performance characteristics of conductive polymers have increased. Numerous companies manufacturing inkjet printable Nano particles have evolved (my database has 32 companies who provide silver Nano particles). Substrates have improved. DuPont Tejin Films have stepped up to provide a material more suited for printed electronics applications. The secondary operations of temperature, vacuum, and light curing effect the substrates. Substrates tend to change shape anisotropicly, making alignment for a second layer process more than difficult. It is best to perform processing on a substrate that doesn't change shape between processing steps. Patterning resolutions have improved, with printing heads touting one (1) Pico liter drop sizes. However, this improvement in drop size equates to slower throughput. Other known printing processes such as offset, gravure, photolithography and transfer printing have demonstrated acceptable resolutions with higher throughput. Slowly, yet step-by-step printed electronic technologies are making it into mainstream production processes. Epson announced their inkjet printed alignment layer in July of 2005. In July of 2006, Epson announced the launch of inkjet printing color filters for Sharp's displays.

PE - Organic Vs Inorganic

The performance levels of organic materials are simply not up to acceptable levels for most applications. There are some applications where comparable high performance is not required, that is where developers should focus their efforts. The performance levels continue to improve every year. However, they are still very susceptible to degradation due to moisture and oxygen exposures. Many encapsulation solutions to these two damaging variables have been tested. To date, the leading solution for a polymer film is developed by IMRE (Singapore). For glass substrates and substrates with limited bending requirements, Corning's Vita™ and Vitex Systems provide a solution. There is one other, but not allowed to disclose at this time.

During the early years of printed electronics researchers struggled with print resolution, chemistries and substrates. Epson's industrial printing research division determined back in 2001 that the secondary processing technologies were just as important and difficult to over come as deposition and associated chemistries. At the beginning of 2006, technical issues around secondary processing were only beginning to be discussed with throughout the PE industry. The industry was trying numerous methods to increase the conductivity of Nano printed materials (and others), while keeping the cure temperature below 200 Deg C. An example of great market timing is the entry time for NovaCentrix. During the 2007 IDTechEx conference on PE in San Francisco, NovaCentrix surfaced. They demonstrated a machine that used photonic curing to anneal (sinter) pre-deposited Nano materials, at room temperature. Just when the industry was beginning to realize secondary processing was not easy, NovaCentrix shows up with a solution. To date, NovaCentrix is supplying high throughput inline production equipment and touting conductivities of 4X bulk for silver and 10X bulk for copper.

Organic materials are liked due to their printing, low temperature processing and flexible characteristics. Inorganic material typically require high secondary processing temperatures (>500 Deg C) and tend not to be susceptible to bending. However, since the material performance of organics is not quite there, companies have begun developing solution process able inorganic materials. Evonik Industries has taken a bottom up approach, increasing the performance while maintaining low annealing temperatures. Inpria has taken a bottom down approach, maintaining performance while lowering the processing temperature. Inpria has developed 325 Deg C processing materials which demonstrate the same material performance as any CVD or PVD process. 325 Deg C is significant because the surface of glass substrates tend to migrate when the processing temperature is over 350 Deg C. Other companies like Kovio have internally developed processing technologies for inorganic materials. Kovio is targeting the RFID market.

RFID

This leads into RFID. Anyone who is in the RFID space knows the revenue is in the back office, not the tag. Avery can supply an RFID tag for US$0.0695. They also claim the delivery of these tags is operating at a loss. A couple years ago, while speaking with Richard Burns, then Director of Proctor & Gamble Company's Global supply chain operations. I asked him, "With all the hype around RFID and claims of the market *taking off* at lower prices, how many could P&G use if they were free?" Richard replied, "Globally? 22 Billion". For those who don't know P&G is the largest consumer of RFID tags. There is a company named Meco located in The Netherlands. At an RFID conference in Florida, Meco demonstrated their RFID machine which could print RFID tags, attach the chip, program the chip and validate the chip in a single process (one machine). The machine's throughput was described at being capable of producing 1 billion tags per month. The cost of the machine was verbally quoted at US$500,000. Two machines can produce P&G's global demand in 11 months for a CAPEX of

US$1M. The message here is directed to innovators who are dreaming of a market producing RFID tags. It might be difficult to locate an investor. You'll need a strong value proposition, one that does not include hiring 6 PhDs at US$250K each to perform research. Your competition is Avery who is currently operating at a loss and for US$1M an investor could annually produce 24 billion tags.

On the subject of back office, SensorConnect developed a technology designed to optimize multi-core CPUs. The platform was designed to solve the technical problems associated with having a single system coherently monitor the fine-grain activities of arbitrarily large supply-chains in real-time, and support arbitrarily large populations continuously querying (push or pull) the state and history of such supply-chains in real-time. SensorConnect installed their configuration on a dual quad core system at the University Of Arkansas's RFID test facility to determine the load limit of the architecture. SensorConnect demonstrated a sustainable ingestion rate of 447,000 events per second (uARK IEEE Paper, *SensorConnect Performance and Scalability Experiments*). Previously Oracle and others would limit out in the 4000 events per second range on the same dual quad core architecture. Through additional testing SensorConnect demonstrated true scalability up through 64 dual-core processors. Why is this important? There is a lot of hype around multi-core CPUs. However, with the addition of each core, the system only demonstrates a marginal increase in performance, not a true scale up. SensorConnect's true scalability enables Real-Time processing at an affordable cost (1 server, not 10). SensorConect's technology is currently under the trademark of GroovySQL™. As "Green" remains an issue, in terms of energy reduction, GroovySQL™ is demonstrating a 16X reduction in required power.

Energy

Over the past few years the importance and value of Energy to society has become front page news. This urgency has inspired investors to better support the "Green" initiative. Green originally focused on environmental issues and for the most part still does, but alternative energies tend to fall under the same category. Emerging technologies for fuel cells, batteries, wind turbines and photovoltaics have also fallen into the mainstream for investment.

The industry standard for portable energy is Lithium Ion batteries, globally adopted. Primary issues with lithium ion batteries are the hazardous material and the life. Depending upon the load profile the life of a lithium ion battery averages 2 – 4 years. As you walk through airports and other facilities, you might have observed AED units (Automatic Electronic Defibrillator) within easy access for emergency situations. Most batteries within these units are lithium ion. Due to battery issues, some units have proven to not function when needed. As a result, the FDA has imposed recalls on units to solve this issue.

Most people simply become accustomed to plugging in their notebook or mobile phone at regular intervals. This behavior instills a sense of energy capacity for each user. Emerging technologies leverage this aspect of culture by comparing their energy densities to lithium ion. Most will tout energy densities of 2X or 4X lithium ion, and occasionally more. To a user this means they can go that amount of time longer between charges, if they remain within their established 'use' profile.

Batteries

Advanced lithium ion batteries demonstrate energy densities of 200 Wh/Kg or 560 Wh/L. When an emerging technology boasts energy densities of 2X or 4X Lithium Ion then one can run the numbers and translate to applications. Most mobile phone batteries have about 3 Watt Hours (Wh) of energy. In April of 2008 an industry colleague (James Lupino) contacted me to inform a novel energy storage technology had crossed his radar. It was an energy storage technology that was demonstrating 11200 Wh/L (about 23X Li). When I asked what the technology was? He replied, "Its confidential and you to need to look it". I explained that I was very busy and didn't have time to jump on a plane just because I heard those numbers. When asked if it was solid state, lithium ion, alkaline, fuel cell, or what? James repeated, "It's confidential". I explained to James that from my perspective it sounded like another guy running around claiming he has solved cold fusion. I need something, even if its high-level, to justify looking at it. Reluctantly, the inventor Dr. James Lai scheduled a conference call which only left the same impression, another guy claiming, "I've solved cold fusion!" I had to inform James that without a high level description, it's not worth my time. There are too many innovators in the market making unrealistic performance claims. A couple months pass when James contacts me again and promises that if I visit, he will purchase me a bottle of wine, even if the technology doesn't pass my sniff test. With the gentleman's agreement in place, an NDA was signed with Northern Lights Semiconductor Corporation (NLSC) and their facility was visited. Summary of the visit is that the technology is real; it is repeatable and has been validated at the Taiyo Uden and Lite-On facilities. The technology is very low cost with very high recharge and discharge capabilities. Dr. Lai has an extensive number of patents filed and is in search of funding.

The Paradigm

Occasionally innovators will experience the same dilemma that Dr. Lai is struggling with. My first impression and for the subsequent months was that the numbers are too far out of the 2X or 4X lithium ion range to be taken serious. Its outside my paradigm of feasibility. Dr. Lai offers to potential collaborators the opportunity to test his design with their equipment in their facility. As one who was originally VERY skeptical and subsequently observing his interactions with other organizations, this paradigm issue has proven to be a real problem. Most companies don't take him up on the offer to test the technology, most simply stand firm in their beliefs that its not real and not worth their time. Leveraging industry contacts and explaining the situation, a few companies have begun to take Dr. Lai up on his offer. Of course their paradigms have shifted as mine did. It's not a magic trick. When Dr. Lai's technology achieves full market adoption, technologies like fuel cells will be competing with hydro, wind, solar and nuclear to re-charge Dr. Lai's technology.

Dr. Lai's technology is just one example of many for which I've witnessed this paradigm problem. As a reviewer of emerging technologies one must take a perspective of: The innovator spent a lot of time developing the technology, for some reason, he believes it has value. It is the technology reviewer's responsibility to understand the innovators perceived value, even if the technology has been surpassed by another in the market.

Photovoltaics

TFT, silicon, organic PV, multi-junction, CIGS, and many other technologies are vying for market position. Each of the emerging technologies compares themselves with an efficiency number. Caution must be exercised when comparing numbers, a 1 cm^2 cell tested under ideal laboratory conditions will easily decrease in efficiency by 30% when scaled up. As emerging technologies attempt to enter the market, they need to be aware of changing market conditions. Market attention and pressure is imposed upon currently adopted technologies to reduce their costs and improve efficiency. This results in a moving target for competition. Occasionally at PV conferences, NREL (National Renewable Energy Laboratory) will present a nice graph depicting over time, each technology's achievements in efficiency. One will hear of recorded efficiency records reported from all around the globe. Having them tested under identical conditions increases credibility. To date, records are in the 42% range.

Signage

This sector is comprised of printed signs and digital (LED, LCD, Plasma, eInk, etc.). Occasionally one will observe an LED numeric sign integrated with graphics to advertise items such as the current level of the lottery. Energy remains to be an issue. Burning the LEDs in conjunction with the lamps currently used on billboards cost money and does not constitute a "Green" technology. TRED Displays have developed a novel, low cost, bi-stable, reflective, Green technology targeted at the digital numeric signage market, enabling HybridSigns™. This technology merges digital and paper signage, enabling digital pricing. The signs are networked together without cables and can be changed remotely from the central office using TredLINK™ software. TRED Displays is rapidly gaining global attention.

The technologies discussed are only a few of the many which have been globally reviewed. There are some in biotech which can perform real-time analysis of hybridization, including transient events. In optics, there are ultra low cost transponders which use ink jet printing to collimate the laser, enabling passive alignment of the optics. Ink jet printed cold fire ceramics have proven to enable three-dimensional construction of high frequency circuits, dramatically reducing size and improving performance. An injection moldable concrete is emerging that can provide component geometries with extremely low thermal expansion at a reasonable cost. The list of emerging technologies crossing the path of one analyzing them is extensive. It is also sad to see many fail due to poor alignment of objectives or simply underestimating the requirements to achieve adoption.

Appendix A

Small to Medium Display Market Data

Data reprinted with permission from DisplaySearch, an NPD Company

Application	Units (000s)					Sequential Change				
	Q4'01	Q1'02	Q2'02	Q3'02	Q4'02	Q4'01	Q1'02	Q2'02	Q3'02	Q4'02
Mobile Phone	102,978	88,119	99,681	105,201	109,305	16.2%	-14.4%	13.1%	5.5%	3.9%
PDAs	2,898	3.039	3,465	3,786	3,855	14.5%	4.9%	14.0%	9.3%	1.8%
Automotive Monitors	1,011	1,107	1,470	1,647	1,395	-13.4%	9.5%	32.8%	12.0%	-15.3%
Video Cameras	2,751	3,402	3,870	3,600	3,300	-9.1%	23.7%	13.8%	-7.0%	-8.3%
Digital Still Cameras	4,230	4,110	5,940	8,505	6,495	-17.7%	-2.8%	44.5%	43.2%	-23.6%

Table A.1 Q4'01 – Q4'02 Small/Medium Display Units and Growth by Application

Application	Y/Y Growth				
	Q4'01	Q1'02	Q2'02	Q3'02	Q4'02
Mobile Phone	-16.1%	-0.8%	18.2%	18.7%	6.1%
PDAs	-33.6%	-26.5%	44.9%	49.5%	33.0%
Automotive Monitors	-18.6%	-1.6%	17.2%	41.1%	38.0%
Video Cameras	-13.9%	6.0%	33.1%	18.9%	20.0%
Digital Still Cameras	2.2%	23.5%	41.9%	65.5%	53.5%

Table A.2 Q4'01 – Q4'02 Small/Medium Display Y/Y Unit Growth by Application

Application	ASPs					Sequential Change				
	Q4'01	Q1'02	Q2'02	Q3'02	Q4'02	Q4'01	Q1'02	Q2'02	Q3'02	Q4'02
Mobile Phone	$8.2	$8.7	$9.1	$10.7	$10.9	-13.7%	6.9%	4.5%	17.1%	2.3%
PDAs	$37.8	$40.6	$40.7	$42.3	$42.1	-10.4%	7.6%	0.3%	3.8%	-0.3%
Automotive Monitors	$102.7	$91.6	$90.3	$94.5	$88.9	-10.0%	-10.8%	-1.4%	4.6%	-5.9%
Video Cameras	$25.2	$23.2	$22.3	$23.5	$22.1	-4.0%	-7.9%	-4.1%	5.6%	-5.9%
Digital Still Cameras	$15.9	$14.8	$13.9	$13.4	$13.3	-4.1%	-7.4%	-5.7%	-3.5%	-0.7%

Table A.3 Q4'01 – Q4'02 Small/Medium Displays Volume Weighted ASPs and Change by Application

Application	Y/Y Change				
	Q4'01	Q1'02	Q2'02	Q3'02	Q4'02
Mobile Phone	-12.8%	-14.4%	-2.3%	12.9%	33.8%
PDAs	36.1%	26.3%	0.7%	0.4%	11.6%
Automotive Monitors	-19.0%	-22.8%	-19.1%	-17.2%	-13.5%
Video Cameras	-29.2%	-24.3%	-18.2%	-10.4%	-12.2%
Digital Still Cameras	-20.5%	-19.4%	-17.7%	-19.1%	-16.3%

Table A.4 Q4'01 – Q4'02 Small/Medium Display Y/Y Volume Weighted ASP Changes by Application

Master Technology	Revenue (US$ Millions)							
	1998	1999	2000	2001	2002	2003	2004	2005
a-Si TFT LCD	6382849	13121713	16399142	14176931	21771447	30782134	44185220	50817378
CRT	22649628	21278154	21891456	19688389	19015283	15564215	13234028	9947785
DLP	81582	776812	108742	127039	190504	419143	613274	630256
EINK	0	0	0	0	0	0	3200	13184
EL	141518	141232	135421	251418	180138	144967	140935	108776
FED	0	510	2448	2769	0	0	0	0
HTPS TFT LCD	224660	311767	494698	558447	659940	618850	573680	511477
LCOS	2820	28310	50582	60405	37079	55978	74876	159227
LTPS TFT LCD	128339	303372	432395	483415	783731	2897425	4843779	5477847
OLED	0	15031	19243	51063	112971	227466	330753	518125
PDP	208109	339598	436560	806537	1423522	2816473	4261635	5101338
PMLCD	2986358	3703330	5268703	4224021	3960680	5207277	6297764	6957895
VFD	469606	524965	644529	686217	682858	699320	705490	622453
Total	33275469	39844794	45883919	41116651	48818153	59433248	75264634	79965741

Table A.5 Revenue from shipped display technology

Region/Country	1999	2000	2001	2002	2003	2004	2005	2006	2007	2008
North America	49,459	53,264	65,212	64,593	69,027	87,873	80,270	80,154	83,374	88,685
Latin America	23,909	27,285	33,606	30,464	36,889	52,306	73,911	75,827	80,209	83,165
Japan	39116	45,372	46,629	40,462	52,197	43,754	48,902	56,033	58,479	56,930
Korea	15,519	12,751	12,222	15,400	14,223	15,983	15,972	17,826	17,101	16,494
China	24,206	50,290	62,975	72,536	100,549	107,609	119,610	130,882	160,654	180,997
Asia Others	20,221	30,802	31,559	49,162	65,626	93,379	115,107	133,476	152,789	165,235
West Europe	99,176	134,812	103,402	91,498	94,374	139,652	104,925	107,362	116,725	133,144
East Europe	13,498	25,542	19,030	27,306	31,986	45,535	57,721	60,827	63,799	68,524
Others	6,212	8,361	8,699	13,845	18,046	43,178	44,314	44,597	49,500	50,502
Growth		33.4%	-1.3%	5.7%	19.2%	30.3%	5.0%	7.0%	10.7%	7.8%
Total Demand	291,315	388,478	383,334	482917	482,917	629,269	660,732	706,983	782,631	843,676

Region/Country	1999	2000	2001	2002	2003	2004	2005	2006	2007	2008
North America	53%	49%	48%	42%	42%	49%	43%	41%	41%	43%
Latin America	57%	44%	40%	30%	29%	33%	39%	36%	35%	34%
Japan	72%	71%	64%	51%	61%	50%	54%	61%	63%	61%
Korea	66%	48%	42%	48%	42%	47%	46%	50%	48%	46%
China	56%	59%	43%	35%	37%	34%	33%	33%	38%	40%
Asia Others	50%	49%	36%	40%	41%	44%	43%	42%	42%	41%
West Europe	64%	55%	37%	30%	30%	45%	34%	34%	37%	42%
East Europe	58%	56%	28%	29%	27%	34%	39%	39%	38%	40%
Others	50%	45%	23%	28%	24%	41%	34%	30%	30%	29%
Total	60%	54%	41%	35%	36%	41%	38%	38%	39%	40%

Table A.6 Mobile Phone Demand (000) and % of Subscribers by Region/Country Forecast

Technology	1999	2000	2001	2002	2003	2004	2005	2006	2007	2008
MSTN	290,815	366,915	334,255	335,166	234,763	141,719	71,594	46,905	40,790	38,348
CSTN	500	20,890	31,831	30,343	138,456	252,036	244,038	191,858	163,009	130,425
AMLCD		615	17,120	39,691	109,699	235,085	341,914	461,337	565,059	645,076
PMOLED		58	128	50		371	1,3666	3,855	4,239	3,604
AMOLED				15		57	1,821	3,028	9,534	26,223
Total	291,315	388,478	383,334	405,265	482,917	629,269	660,732	706,984	782,631	843,676

Table A.7 Mobile Phone Shipments by Display Technology Forecast

Supplier	Shipments (1,000 Units/Month)					Share				
	Q4'01	Q1'02	Q2'02	Q3'02	Q4'02	Q4'01	Q1'02	Q2'02	Q3'02	Q4'02
AU Optronics										
Casio	150	215	225	120	80	0.4%	0.7%	0.7%	0.3%	0.2%
Epson	9,590	7,135	7,035	7,750	8,120	27.9%	24.3%	21.2%	22.1%	22.3%
Hitachi	280	520	485	470	740	0.8%	1.8%	1.5%	1.3%	2.0%
NEC	415	330	215	450	240	1.2%	1.1%	0.6%	1.3%	0.7%
Matsushita	378	386				1.1%	1.3%			
Samsung	90	280	400	700	830	0.3%	1.0%	1.2%	2.0%	2.3%
Sanyo			5	55	85			0.0%	0.2%	0.2%
Sharp	793	888	1,270	1,380	1,590	2.3%	3.0%	3.8%	3.9%	4.4%
ST-LCD										
Toshiba	470	354				1.4%	1.2%			
TMDisplay			1,047	1,172	1,900			3.2%	3.3%	5.2%
Optrex	2,470	1,820	1,380	1,480	1,580	7.2%	6.2%	4.2%	4.2%	4.3%
Philips	6,160	5,015	7,225	6,175	7,350	17.9%	17.1%	21.7%	17.6%	20.2%
Picvue	120	75	100	200	250	0.3%	0.3%	0.3%	0.6%	0.7%
Samsung SDI	6,780	6,820	7,890	8,950	7,945	19.8%	23.2%	23.7%	25.5%	21.8%
SII	2,420	2,370	2,730	2,920	2,570	7.1%	8.1%	8.2%	8.3%	7.1%
Three-Five	750	300				2.2%	1.0%			
Wintek	1,100	1,100	1,120	1,300	1,250	3.2%	3.7%	3.4%	3.7%	3.4%
Others	2,360	1,765	2,100	1,945	1,905	6.9%	6.0%	6.3%	5.5%	5.2%
Total	34,326	29,373	33,227	35,067	36,435	100%	100%	100%	100%	100%

Table A.8 Q4'01 – Q4'02 Mobile Phone Display Module Shipments and Share by Supplier

Supplier	Revenues ($US Millions/Quarter)					Share				
	Q4'01	Q1'02	Q2'02	Q3'02	Q4'02	Q4'01	Q1'02	Q2'02	Q3'02	Q4'02
AU Optronics										
Casio	$14.2	$17.1	$19.3	$10.3	$5.3	1.7%	2.2%	2.1%	0.9%	0.4%
Epson	$234.1	$177.4	$207.0	$291.0	$276.3	27.8%	23.1%	22.8%	25.9%	23.1%
Hitachi	$30.3	$52.8	$45.6	$44.3	$65.9	3.6%	6.9%	5.0%	3.9%	5.5%
NEC	$47.4	$33.6	$21.2	$43.9	$20.9	5.6%	4.4%	2.3%	3.9%	1.7%
Matsushita	$16.1	$21.0				1.9%	2.7%			
Samsung	$10.0	$28.5	$37.8	$65.8	$71.6	1.2%	3.7%	4.2%	5.9%	6.0%
Sanyo			$0.8	$5.8	$8.0			0.1%	0.5%	0.7%
Sharp	$57.7	$72.8	$103.8	$105.0	$107.8	6.9%	9.5%	11.4%	9.3%	9.0%
ST-LCD										
Toshiba	$52.3	$36.1				6.2%	4.7%			
TMDisplay			$97.3	$109.2	$162.9			10.7%	9.7%	13.6%
Optrex	$46.7	$39.6	$25.7	$40.9	$46.0	5.5%	5.1%	2.8%	3.6%	3.9%
Philips	$99.9	$78.1	$109.2	$109.4	$165.6	11.9%	10.2%	12.0%	9.7%	13.9%
Picvue	$1.9	$1.2	$1.5	$3.2	$3.8	0.2%	0.2%	0.2%	0.3%	0.3%
Samsung SDI	$114.1	$112.6	$129.5	$167.6	$154.1	13.6%	14.6%	14.2%	14.9%	12.9%
SII	$48.2	$48.6	$61.5	$76.0	$58.6	5.7%	6.3%	6.8%	6.8%	4.9%
Three-Five	$12.2	$4.7				1.4%	0.6%			
Wintek	$17.8	$17.1	$16.9	$20.7	$19.1	2.1%	2.2%	1.9%	1.8%	1.6%
Others	$38.3	$27.9	$32.1	$30.9	$29.1	4.5%	3.6%	3.5%	2.7%	2.4%
Total	$841.2	$769.1	$909.2	$1,124.0	$1194.9	100%	100%	100%	100%	100%

Table A.9 Q4'01 – Q4'02 Mobile Phone Display Module Revenues and Share by Supplier

Technology / Brands	1,000 Units/Month					Share				
	AMLCD	CSTN	MSTN	OLED	LCD Total	AMLCD	CSTN	MSTN	OLED	LCD Total
Alcatel			900		900			100%		100%
Fujitsu	342				342	100%				100%
Kyocera	80	50	800		930	8.6%	5.4%	86.0%		100%
LG InfoComm	55	130	805		990	5.6%	13.1%	81.3%		100%
Mitsubishi	190	20	200		410	46.3%	4.9%	48.8%		100%
Motorola		130	3,675		3,805		3.4%	96.6%		100%
NEC	485	110			595	81.5%	18.5%			100%
Nokia	180	70	11,700		11,950	1.5%	0.6%	97.9%		100%
Panasonic	395	675	150		1,220	32.4%	55.3%	12.3%		100%
Samsung	320	350	2,360		3,030	10.6%	11.6%	77.9%		100%
Sanyo	140	145	120	5	410	34.1%	35.4%	29.3%	1.2%	100%
Sharp	580				580	100%				100%
Siemens			2,650		2,650			100%		100%
Sony Ericson	220	355	1,900		2,475	8.9%	14.3%	76.8%		100%
Toshiba	430		230		660	65.2%		34.8%		100%
Others	250	250	1,780		2,280	11.0%	11.0%	78.1%		100%
Total	3,667	2,285	27,270	5	33,227	11.0%	6.9%	82.1%	0.0%	100%

Table A.10 Q4'01 – Q2'02 Mobile Phone Display Module Value Chain by Technology and Share

Set Maker	China		Japan			Korea			Western Europe									USA			
	China Mobile	China Uni Com	NTT Docomo	Vodafone	AU	SK Telecom	KTF	LG Telecom	TIM (Italy)	Vodafone (Italy)	Vodafone (Germany)	T-Mobile (Germany)	Orange (France)	SFR (France)	O2 (UK)	Vodafone (UK)	BT (UK)	Verizon	Cingular	AT&T	Sprint PCS
Alcatel	○									○			○								
Fujitsu			○																		
Kyocera					○													○			
LG						○	○	○				○	○			○		○	○	○	○
Mitsubishi			○	○																	
Motorola		○				○	○		○	○	○	○	○	○	○	○	○	○	○	○	○
NEC	○		□	○																○	
Nokia	○	○							□	○	□	□	○	○	○	□	○	□	○	○	○
Panasonic	○	○	□							○	○	○	○	○		○	○				
Samsung	○	○				□	○		○	○	○	○	○	○	○	○	○	○	○	○	□
Sanyo				○	□	○															○
Sharp			○	□						○	○	○		○		○					
Siemens	○	○							○	○	□	□	○		○	○	○		○	○	
SonyEricsson	○	○	○		○				○	○	○	○	○	○	○	○	○		○	○	

Table A.11 Service Provider and Set Maker Relationship

Appendix **B**

Emerging Technology Primer

Recommend to be used during initial communications to determine interest with a potential collaboration partner or investor. Next step, use comprehensive presentation.

Problem:

What is the industry pain?

Company Logo

Solution:

What is the company solution/value proposition (bullet points)?

Company Information: Date Founded

Name:
Address:

Phone:

Contact Name:

Summary:

(Brief overview of company.)

Technology

High level view of technology, can include graphics

Market Analysis

Industry leaders know the market; keep simple to demonstrate knowledge. 3-4 sentences with key words: size, sectors applications, etc.

Development Status

Identify level of status with number of patents, proof of concept, collaborations, etc.

Competition

What is the competition and what is the industry technology migration?

Business Model

Clearly define, don't be vague

Note: This section can be reserved for a later meeting.

Appendix C

Representative sample of Kaizen Continuous Improvement Program. Courtesy of Epson Portland Inc.

SUMMARY

The purpose of the Kaizen Continuous Improvement Program is to provide a formal recognition and reward system to promote continuous improvement activities (Kaizen) through employee creativity. Through this program, we intend to encourage employees to:

1. Initiate improvements in their work and make corrective actions, and/or preventive actions to processes.
2. Be recognized for continuous improvements they have achieved.
3. Provide an opportunity for employees to reach higher levels of ability.
4. Increase morale in the work areas.
5. Strengthen manufacturing capabilities through improvements in Safety, Quality, Cost, Delivery, and Environmental aspects ("SQCDE").

RESPONSIBILITY

Policy Revisions – Director of Administration
Enforces/Promotes – Manager of Customer Satisfaction Quality Assurance (CSQA)
Applies to – Employees and Temporary Contractors

DOCUMENTATION

Kaizen Continuous Improvement form

WHO CAN PARTICIPATE

Anyone can participate. All employees, both regular and temporary, may submit Kaizens in order to document their individual accomplishments or the accomplishments of a team of up to ten members. Employees may also use the Kaizen Continuous Improvement form to submit suggestions for improvements they are not able to implement themselves.

AREAS OR TARGETS FOR IMPROVEMENT

Virtually any work-related idea that results in improving, correcting or preventing such things as:

- A work process or task.
- A safety or environmental issue.
- An increase in efficiency or productivity.
- An increase in quality.
- A cost reduction or cost avoidance.

Process - A

1. Implement a corrective or preventive action or work improvement.
2. Complete a Kaizen Continuous Improvement form, describing what was accomplished and estimating the value of the improvement.

3. Submit the Kaizen form to the next level of direct management (supervisor, assistant manager, manager or director).
4. Based upon the rating, the appropriate signatures are obtained. Kaizens resulting in an "encouragement" or "effort" reward require supervisor signature. "Achievement" awards require the manager's signature.
5. Submit the completed form to the Manager of CSQA .
6. The Manager of CSQA will record points and the award.
7. All payouts (except for temporaries) will be processed through Payroll.
8. Payouts will be monthly, timed to coincide with the All Employee Meetings. The minimum payout will be $10. Amounts less than that will be accumulated until they total $10 for the employee. Temporary employees will receive a gift card.
9. If an employee is owed a balance of less than $10 at the end of a calendar year, the company will compensate the employee with a gift of similar value and set the balance to zero.

PROCESS - B

1. Complete a Kaizen Continuous Improvement form, describing the suggested corrective or preventive action or work improvement.
2. Submit the Kaizen to the CSQA department.
3. CSQA management will review the suggestion and discuss the feasibility of implementation with the necessary departments.
4. If implemented, the originator will receive partial points for submitting the idea.

CLASSIFICATION AND AMOUNT OF AWARDS FOR EACH KAIZEN SUBMITTED

Classification	Significance	Points	Award
Encouragement	Small	2	$10.00
Effort	Medium or Good	6	$20.00
Achievement	Large or outstanding	20	$60.00

PRESENTATION / PROMOTION

In order to share ideas and promote the Kaizen Program, Kaizens may appear on company bulletin boards, be featured in the company newsletter, or be presented during All Employee or Quality Meetings.

KAIZEN MONTH

Prior to the start of Kaizen Month each July, CSQA will announce various activities to help promote participation, increase awareness of the program and showcase examples of Kaizen improvements. Promotional prizes and/or competitions may be included.

QUALITY MONTH PRESENTATIONS

CSQA management may select several individual Kaizens to be presented at the presentations, in addition to team Kaizens. Each Kaizen presented will be eligible for an additional $50 award. Some employees may be selected to present their Kaizens at regional quality presentations and possibly the worldwide team competitions at company headquarters.

Since Quality Month is in November, it will be necessary to begin refining presentations during the summer prior in order to be prepared for Quality Month presentations.

YEARLY AWARDS (FISCAL YEAR, FROM APRIL TO MARCH)

Soon after fiscal year end, the ten employees from the entire company who have earned the greatest number of points (minimum of 20 points to qualify) and who have implemented four or more ideas in one year will each be awarded $50.

The employee with the highest number of implemented Kaizens will be awarded $25.

CUMULATIVE AWARDS

Kaizen contributors will receive a $50.00 award for every 100 points earned (no time restrictions).

EXPENSE RESPONSIBILITY

Award expenses will be budgeted and charged directly to the employee's department from which the idea(s) originated.

GUIDELINES FOR MANAGER REVIEW OF IDEA RESULTS

Using guidelines on the back of the Kaizen form, evaluate the improvement.
Standard Hourly Rate = $15/hour (includes benefits)

EXAMPLE CALCULATION METHODS

1. Labor Improvement: "Before Time" (time required before improvement) minus "After Time" (time required after improvement) equals Time Saved.

Step 1: Man hours saved per day	Step 2: Man hours saved per month	Step 3: Dollar savings per month
10.0 hours/day ("before" time)	2.5 hours/day	52.5 hours/month
-7.5 hours/day ("after" time)	x21.0 days/month	x $15.00/hour
2.5 hours/day	52.5 hours/month	$787.50/month(labor improvement)

2. Parts or Material Reduction:

Step 1: Assumptions	Step 2: Savings per day	Step 3: Savings per month
(Improvement) Decreased number of damaged connectors by 3%	(3,000 Units/day x 0.03)	$9.90 savings/day
Cost per connector: $0.011	x $0.11/Unit	x 21 days/month
Daily production = 3,000 Units per day	$ 9.90 savings/day	$207/month (parts/material reduction)

References

Preface

1. A 2007 study of 539 angels involved with groups, sponsored by the Angel Capital Education Foundation and the Kauffman Foundation, reported in the Wall Street Journal

Chapter 1 Introduction

1. Penny Le Couteur & Jay Burreson (2003) *Napoleon's Buttons*.
2. Wall Street Journal Classroom Edition, "Adoption rates of various communication technologies" Original data was lost during September 11, 2001 attack on World Trade Center.
3. Krusic PJ, Wasserman E, Keizer PN, Morton JR, Preston KF, Science 1991; 254: 1183.
4. Chiang LY, Lu FJ, Lin JT, J. Chem. Soc., Chem. Commun. 1995: 1283.
5. Dugan LL, Gabrielsen JK, Yu SP, Lin TS, Choi DW, Neurobiol Dis. 1996; 3(2): 129.
6. Okuda K, Mashino T, Hirobe M, Bioorg. Med. Chem. Lett. 1996; 6: 539.
7. Imai, Masaaki (1986) *Kaizen: the key to Japan's competitive success*, McGraw-Hill, Inc.

Chapter 2 Emerging Technologies

1. Interview with Dr. Jeremy Burroughes, accredited inventor of Polymer Organic Light Emitting Diode technology.
2. Gartner Hype Cycle, http://www.gartner.com/pages/story.php.id.8795.s.8.jsp
3. Die Another Day (2002) The 20th spy film in the James Bond series. Produced by Barbara Broccoli, Michael G Wilson, and Anthony Waye.
4. Interview with Dr. Marc Pauchard of ILFord Imaging, Fribourg, Switzerland
5. Art Fry and the invention of Post-it® Notes http://www.3m.com/about3m/pioneers/fry.jhtml
6. Post-It® is a registered trademark of 3M
7. Arthur Fry on Lemelson Website http://www.inventionatplay.org/inventors_fry2.html
8. Planar financials reported by the Securities and Exchange Commission
9. Siegel, Donald S. et al. (2004) *Toward a model of the effective transfer of scientific knowledge from academicians to practitioners: qualitative evidence from the commercialization of university technologies*. J. Eng. Technol. Manage. 21 (2004) 115–142
10. Mejia, Luis R. (1998) *A Brief Look at a Market-Driven Approach to University Technology Transfer: One Model for a Rapidly Changing Global Economy,* Technological Forecasting and Social Change 57, 233-235

Chapter 3 Technology Development

1. Fyfe, David (Presentation) *Invention to Commercialization*, Lux Executive Summit, October 2007
2. Agency for Science, Technology and Research (A*STAR) Singapore
3. University of Manchester UK Intellectual Property Limited (UMIP)
4. Moore,Geoffrey A. (1999) *Crossing the Chasm, Marketing and Selling High-Tech Products to Mainstream Customer (revised edition)*, HarperCollins Publishers
5. Getting the most out of your product development process. Harvard Business Review March–April 1996
6. *Technology Development Stage-Gate Process* Product Development Institute, Inc., (2007) (www.prod-dev.com)
7. Project Management Institute (2004) *A guide to the Project Management Body of Knowledge, 3rd Edition* (PMBOK®) www.pmi.org
8. Montgomery, Douglas C (2004) *Design And Analysis Of Experiments* John Wiley & Sons
9. Taguchi, Genichi and Yokoyama, Yoshiko (1993) *Taguchi Methods: Design of Experiments (Taguchi Methods Series),* Amer Supplier Inst
10. Box, George E.P. et al (1978) *Statistics for Experimenters, an Introduction to Design, Data Analysis, and Model Building* John Wiley & Sons
11. Moore,Geoffrey A. (1999) *Crossing the Chasm, Marketing and Selling High-Tech Products to Mainstream Customer (revised edition)*, HarperCollins Publishers
12. Carlson, Curtis R. and Wilmot, William W. (2006) *Innovation, The five disciplines for Creating What Customers Want* Crown Business
13. International Project Management Association www.ipma.ch

Chapter 4 Strategic Marketing

1. Grant, Robert M. *Contemporary Strategy Analysis* 3rd Edition, 1998, Blackwell Publishers Inc.
2. Hatton, Angela (2000), *The definitive guide to marketing planning*, Pearson Education Ltd.
3. Doyle, Peter (2000) *Value-Based Marketing*, John Wiley & Sons Ltd.
4. RoHS Regulations, European Directive 2002/95/EEC
5. http://www.fda.gov/oc/initiatives/counterfeit/report02_04.html
6. Wal-Mart Mandate (June 10, 2003) Linda Dillman, CIO Wal-Mart, *Mandates that their top 100 suppliers will be required to use radio frequency identification(RFID) tags on their cases and pallets by January 2005*. RFID Journal June 11, 2003
7. U.S. Food and Drug Administration (November 2004) *Radiofrequency Identification Studies and Pilot Programs for Drugs http://www.fda.gov/oc/initiatives/counterfeit/rfid_cpg.html*
8. Porter, Michael (1998) *Competitive Strategy: Techniques for analyzing Industries and Competitors*, The Free Press
9. The Boston Consulting Group www.bcg.com
10. Koller, Tim et al (2005) *Valuation, Measuring and Managing the Value of Companies*. McKinsey & Company
11. Slywotzky, Adrian and Morrison, David (2001) *The Profit Zone*. Three Rivers Press
12. Blank, Steven Gary (2006) *The Four Steps to the Epiphany, 3rd Edition* Cafepress.com
13. Doyle, Perter and Stern, Phillip (2006) *Marketing Management and Strategy, 4th* Edition, Pearson Education Limited

Chapter 5 Business Development

1. Gan Chee Eng, Vice President (Greater China Region) of Amway China Company Limited, Cit.
2. Alticor Press release October 21, 2004 *Alticor Sales Surge Past $6 Billion* http://amway.com/en/General/newsroom.aspx
3. Tusa, Stephan C., Jr et al (2006) *Report: Electrical Equipment and Multi-Industry* J.P. Morgan Securities Inc.
4. Harrop, Peter (2006) *Article: The Price-Sensitivity Curve for RFID* IDTechEx http://www.idtechex.com/products/en/articles/00000488.asp
5. Das, Raghu (2006) *Article: RFID – Big Orders Imminent* IDTechEx http://www.idtechex.com/products/en/articles/00000507.asp
6. Wal-Mart Mandate (June 10, 2003) Linda Dillman, CIO Wal-Mart, *Mandates that their top 100 suppliers will be required to use radio frequency identification(RFID) tags on their cases and pallets by January 2005*. RFID Journal June 11, 2003
7. U.S. Food and Drug Administration (November 2004) *Radiofrequency Identification Studies and Pilot Programs for Drugs http://www.fda.gov/oc/initiatives/counterfeit/rfid_cpg.html*
8. Cambrios Technologies Corporation www.cambrios.com
9. http://www.earthsky.org/teachers/articles.php?id=3 (Seashells, August 2000 by Shireen Gonzaga and Marc Airhart)
10. Curtis, Valerie (2002) *Lessons from building Public-Private Partnerships for Washing Hands with Soap* Health in Your Hands (www.globalhandwashing.org)
11. Curits, Valerie October 2002, "Health in your hands: lessons from Building Public-Private Partnerships for Washing Hands with Soap."
12. www.globalhandwashing.org/publications/lessons_learntpart1.htm
13. C.K. Prahalad (2004) *The Fortune at the Bottom of the Pyramid*. Wharton School Publishing
14. Howse, Derek (1989) *Neville, Maskelyne, the Seamen's Astronomer* Cambridge Universtiy Press
15. Taylor, E.G.R. (1957) The Haven-Finding Art: *A History of Navigation from Odysseus to Captain Cook*, Abelard-Schuman Ltd
16. Helden, Albert Van (1995) *Longitude at Sea*, The Galileo Project
17. Gladwell, Malcolm (2002) *The Tipping Point: How Little Things Can Make a Big Difference* Back Bay Books
18. Day, George S, et al, (2004) Wharton on Managing Emerging Technologies, John Wiley & Sons
19. SEC Filings March 2000 through March 2004, Telecommunications Market
20. Blank, Steven Gary (2006) *The Four Steps to the Epiphany, 3rd Edition* Cafepress.com
21. Carlson, Curtis R. and Wilmot, William W. (2006) *Innovation, The five disciplines for Creating What Customers Want* Crown Business

22. Fyfe, David (Presentation) *Invention to Commercialization*, Lux Executive Summit, October 2007
23. Intel Capital CEO Summit and Technology Days published with permission
24. Kuglin, Fred A. (2002) *Building, Leading and Managing Strategic Alliances* Amacom
25. Hart, Stuart L. (2005) *Capitalism at the Crossroads* Wharton School Publishing
26. Dupont acquires Uniax, www.dupont.com/corp/news/releases/2000/nr03_16_00.html
27. Sanyo Presentation at FPD in Yokohama, 2002
28. The Rise and Rise of the Redmond Empire
29. http://www.wired.com/wired/archive/6.12/redmond.html

Chapter 6 Legal Protection

1. The Road To WiMAX How Intel's Sean Maloney shepherded through the technology that's poised to rewrite the rules of wireless, BusinessWeek Sept 3, 2007
2. Kumar, N Senthil (2008) Understanding Patent Rights, Ezinearticles.com
3. Breitzman, Anthony (2001) *Patent count vs GDP by inventor country,* Reproduced with permission.
4. Display Market data provided by Display Search, An NPD Company. Reproduced with permission
5. Project Support Desk (2002) Manageming Intellectual Property for Maximum Value, Corporate Executive Board
6. Evans, Paula C. and Smith, Richard B. (2002) *Intellectual Property Rights in Collaborations* Bio Tech International, Volume 14, Page 8-9

Chapter 7 Financial

1. DePamphilis, Donald (2001) *Mergers Acquisitions and other Restructuring Activities* Academic Press
2. JP Morgan Securities Report September 22, 2006, Product ID Competitive Financial Data Hutchison and Mason PLLC (2001) *Structuring Venture Capital Investments* Triangle TechJournal
3. Doyle, Peter (1998) *Marketing Management and Strategy* 2nd Edition, Prentice Hall Europe
4. Doyle, Peter (2000) *Value-Based Marketing*, John Wiley & Sons Ltd.
5. Grant, Robert M. *Contemporary Strategy Analysis* 3rd Edition, 1998, Blackwell Publishers Inc.
6. Doyle, Perter and Stern, Phillip (2006) *Marketing Management and Strategy, 4th* Edition, Pearson Education Limited
7. Hatton, Angela (2000), *The definitive guide to marketing planning*, Pearson Education Ltd.
8. Rankine, Denzil et al (2003) *Due Diligence definitive steps to successful business combinations.* Prentice Hall
9. Koller, Tim et al (2005) *Valuation, Measuring and Managing the Value of Companies*. McKinsey & Company
10. Hitchner, James R. (2006) *Financial Valuation, Second Edition* John Wiley & Sons
11. Lightbody, Heather (2006) *Understanding Company Valuation and Acquisitions – a modular program* London Stock Exchange
12. Sahlman, William A. (2004) *The Basic Venture Capital Formula* Harvard Business School Publishing
13. Villalobos, Luis *Exit Ratios and their Implications for Venture Investment Valuations* http://www.angelcapitalassociation.org/dir_about/news_detail.aspx?id=79
14. Bapat, Amit (2004) *How to value startups and emerging companies* Oregon Health Sciences University
15. Cunningham, Jean and Fiume, Orest (2003) *Real Numbers*, Partners Pub Group, Inc.

Chapter 8 Organizational Management

1. Aaker, David (2004) *Brand Portfolio Strategy*. Free Press
2. U.S. Food and Drug Administration (November 2004) *Radiofrequency Identification Studies and Pilot Programs for Drugs http://www.fda.gov/oc/initiatives/counterfeit/rfid_cpg.html*
3. Egan, Colin (1995) *Creating Organizational Advantage* Butterworth-Heinemann
4. Blank, Steven Gary (2006) *The Four Steps to the Epiphany, 3rd Edition* Cafepress.com
5. FUD was first defined by Gene Amdahl after he left IBM to found his own company, Amdahl Corp.: "FUD is the fear, uncertainty, and doubt that IBM sales people instill in the minds of potential customers who might be considering Amdahl products." Consult Wikipedia at http://en.wikipedia.org/wiki/FUD

6. Carlson, Curtis R. and Wilmot, William W. (2006) *Innovation, The five disciplines for Creating What Customers Want* Crown Business
7. Kohn, Alfie (1990) *The Brighter Side of Human Nature: Altruism and Empathy in Everyday Life* Basic Books
8. http://changingminds.org/disciplines/leadership/articles/manager_leader.htm
9. Kouzes and Posner, *The Leadership Challenge*, Jossey Bass, 2002
10. http://www.1000ventures.com/business_guide/crosscuttings/leadership_vs_mgmt.html
11. Haas and Madison (2007) *Success on Purpose* Think TQ Inc.
12. Robert Slater (2003) *29 Leadership Secrets from Jack Welch.* McGraw Hill
13. Grant, Robert M. *Contemporary Strategy Analysis* 3rd Edition, 1998, Blackwell Publishers Inc.
14. Howland, Jim (1982) *Jim's Little Yellow Book* Ch2m Hill Corp.
15. 2007 Ch2m Hill Corporate annual report, *Just the Facts*
16. 2007 McDonald's annual report
17. www.pmi.org
18. www.ipmi.org.cn
19. www.PgMP.com
20. Association of Professionals in Business Management, www.apbm.com
21. Vallabhaneni, S. Rao (2008) *Corporate Management, Governance, and Ethics BEST PRACTICES* John Wiley & Sons
22. Gladwell, Malcolm (2002) *The Tipping Point: How Little Things Can Make a Big Difference* Back Bay Books
23. Collins, James C and Porris, Jerry I (1994) *Built to Last – Successful habits of visionary companies* Harper Collins Publishers
24. Hatton, Angela (2000), *The definitive guide to marketing planning*, Pearson Education Ltd.

Chapter 9 Corporate Perspectives

1. C.K Moore,Geoffrey A. (1999) *Crossing the Chasm, Marketing and Selling High-Tech Products to Mainstream Customer (revised edition)*, HarperCollins Publishers
2. Prahalad (2004) *The Fortune at the Bottom of the Pyramid.* Wharton School Publishing
3. Cooper, Robert *et al.* (2001) *Portfolio Management for New Products.* 2nd Edition, Perseus
4. Aaker, David (2004) *Brand Portfolio Strategy.* Free Press
5. Little, Arthur D (1981) *The Strategic Management of Technology*, European Management Forum
6. Christensen, Clayton M. (2003) *The Innovator's Dilemma.* Harper Business Essentials
7. Slywotzky, Adrian J. (1996) *Value Migration.* Harvard Business School
8. Slywotzky, Adrian and Morrison, David (2001) *The Profit Zone.* Three Rivers Press
9. C.K. Prahalad (2004) *The Fortune at the Bottom of the Pyramid.* Wharton School Publishing
10. Robert Slater (2003) *29 Leadership Secrets from Jack Welch.* McGraw Hill
11. Hussey, D. E. *Portfolio analysis: practical experience with the directional policy matrix*, Long Range Planning, August 1978 pp 78-89.

Chapter 10 Evaluating Emerging Technologies

1. Process derived by author from many years of managing and evaluating emerging technologies.
2. Harrop, Peter (2006) *Article: The Price-Sensitivity Curve for RFID* IDTechEx http://www.idtechex.com/products/en/articles/00000488.asp
3. Das, Raghu (2006) *Article: RFID – Big Orders Imminent* IDTechEx http://www.idtechex.com/products/en/articles/00000507.asp

Chapter 11 Application Poly-OLED

1. Industry Press releases (www.oled-info.com/history)

   1996 - CDT gives world's first public demonstration of Light Emitting Polymer devices
   1997 - UDC Demonstrates Flexible Flat Panel Display Technology
   1997 - Pioneer Electronic Produces EL Display with 260,000 Colors
   1998 - Kodak, Sanyo Show Full-Color Active Matrix Organic Display; First Color OEL Display Increases Threat to LCD Display Dominance

2000 - Ritek plans to mass produce OLED
2000 - Toshiba Corp. plans to produce (organic EL) panels in 2001
2000 - Motorola Grants OLED Technology Rights To Universal Display And Takes Equity Position
2000 - UDC and PPG Industries Form Strategic Alliance for Development & Supply of Chemicals for OLED Manufacturers
2000 - Sanyo Electric to start mass production of color organic EL panels in 2001
2000 - NEC, Samsung To Develop Organic Wireless Displays
2000 - LG Electronics develops organic EL displays for mobile gadgets

2001 Industry Press Releases
February - Sony Develops World's Largest Full Color OLED (13 inches diagonally with a resolution of 800x600 pixels.)
April - Universal Display Corporation and Sony Corporation Announce Joint Development Agreement Aimed at OLED Television Monitors.
April - Samsung Exhibits Color Organic EL LCD Panel for Mobile Phones at CeBIT (132 by 162 pixels).
May - Toshiba Develops World's First 260,000-Color Polymer OLED.
August - eMagin's OLED Microdisplay Selected by Air Force for F15E Aircraft.
October - Sony Demonstrates 13-Inch, Full-Color OLED Prototype.
October - Universal Display Corporation and Samsung SDI Announce Key Joint Development Agreement.
October - Universal Display Reports New Red Phosphorescent OLED Materials, Significant Power Efficiency Advances.
October - Samsung SDI Develops World's Largest Organic Display Panel (15.1 inch).
November - RiTdisplay opens new color OLED/PLED factory.

2002 Industry Press Releases
February - Samsung SDI develops 2.2-inch AM OLED for mobile phones.
April - Philips Announces Industry's First Volume Shipments of Polymer-Based OLED Modules.
May - Kodak Announces Availability of Evaluation Kit For Active-Matrix OLED Display.
May - RiTdisplay reportedly receives orders for over one million mobile phone OLEDs.
June - AU Optronics develops world's first OLED prototype combining a-Si TFT LCD technology.
June - Epson, CDT form polymer OLED production venture.
November - Pioneer Supplies OLED to LG Electronics for Cell Phones.
December - DuPont Displays and Universal Display Corporation Form Strategic Alliance to Develop Next Generation Displays Combining Aspects of Small Molecule and Polymer OLEDs.

2003 Industry Press Releases
January - Philips invents EL material to generate both red and green light.
January - Kodak, Sanyo invest in active-matrix OLED production.
January - Sanyo Unveils Mobile Phone with Organic EL Panel Designed for KDDI, announces plans to build full-color OLED production line.
February - RiTdisplay lands mobile phone OLED orders from Samsung and LGE.
February - Kodak Licenses Samsung NEC Mobile Display to Manufacture Passive-Matrix OLED Displays.
March 03 2003 Kodak introduces EasyShare LS633. The world's first digital camera with an OLED display. The camera has 3.1Mpixels and 3X optical zoom. 2.2" OLED display with 512 x 218 pixels
March - IDTech develops 20-inch full-color OLED display.
April - DuPont forges 'Olight' brand for emerging OLEDs.
April - Sony testing 24-inch OLED screen.
May - AUO and UDC develop 4-inch, a-Si TFT backplane-based, red phosphorescent AMOLED.
May - Samsung NEC develops 65,000-color PM OLED mobile phone display.
May - Sony demonstrated a 24.2 inch OLED panel.
June - ERSO and Windell develop 10-inch LTPS TFT technology-based AM OLED display.
June - Sony to invest nine billion yen to build OLED production line.
September - AU Optronics has showcased its 1.93" AMOLED for Mobile Phone.
September - Princeton electrical engineers have invented a technique for making Organic Solar Cells that Could Lead To Widespread Use Of Solar Power.

October - Sanyo Unveils Long-Lasting QVGA Organic EL Panel for Cell Phones.
October - Univision begins OLED production.
November - Tohoku Pioneer Corporation to Be First Manufacturer to Use Universal Display Corporation's PHOLED Material in a Commercial OLED Display.
November - Univision showcases 65,000-color mobile phone OLED sub-display.
December - Univision reaches monthly capacity of 6,000 OLED substrates.
December - Casio Ventures into Organic EL Panels Driven by Amorphous Si TFT.

2006 Industry Press Releases
18 April 2006 Kodak Licenses OLED Technology to Univision Technology Inc. of Taiwan WEBWIRE
14 March 2006, Epson Develops the World's First Print Head Using an OLED Light Source, Seiko Epson Corp.
01 January 2006 Kodak Broadens its Participation in OLED Technology, ends OLED joint venture with Sanyo, Kodak Corporation

2007 Industry Press Releases
02 August 2007, Sumitomo Chemical to Make CDT Wholly-owned Subsidiary, Masao Oonishi, Nikkei Microdevices
14 October 2007, Seiko Epson expanded OLED lifetime, to start making OLED panels (8" to 21"), The Electronic Times
29 October 2007 Samsung's OLED roadmap - 21" monitors by 2009, 42" Full HD TV by 2010, engadget Donald Melanson

2. Cambridge Display Technologies Press Releases

2000
21 January 2000, CDT Expands Light Emitting Polymer (LEP)
09 March 2000, CDT Demonstrates Success of Partnership...
09 March 2000, CDT and Bayer AG Strike Joint Development
09 March 2000, CDT Begins Operation of $25 Million...(US Version)
15 March 2000, DuPont i-Technologies acquires UNIAX Corporation (www.dupont.com)
21 March 2000, Giant OEM Supplier to Develop Computer Displays
10 April 2000, CDT Names Veteran Global Business Leader
09 May 2000, CDT The Next Stage of Development
25 May 2000, CDT is Case Study for University Exploitation
22 June 2000, Seiko-Epson & CDT Agree Terms of Joint License
22 June 2000, CDT Scales Up LEP Material Production
22 June 2000, Seiko-Epson and CDT Develop World's First
04 December 2000, CDT Commits $25 Million for LEP Manufacturing
06 December 2000, CDT Appoints Stewart Hough as Vice-President
06 December 2000, CDT Appoints Keith Bergelt as Vice-President

2001
15 May 2001, CDT Joins US Display Consortium's Governing
17 May 2001, CDT Starts First Phase of New Development
22 May 2001, CDT Grants LEP License to OSRAM
30 May 2001, CDT and Tokki Corporation to Develop
28 August 2001, CDT Selected to Develop Prototype LEP Display
01 October 2001, CDT Grants LEP License to Microemissive Display
13 November 2001, Secret Behind High Efficiency of LEP Disclosed
14 November 2001, CDT Names Jeremy Burroughes Chief Technology
15 November 2001, CDT and Dow Reach Agreement on LEP
26 November 2001, CDT Signs License and Technical Assistance

2002
02 January 2002, Covion and Cambridge Display Technology

29 January 2002, CDT begins operation of $25 Million
15 February 2002, STMicroelectronics and CDT Sign Agreement
04 March 2002, Plastic Logic and Cambridge Display Technology
21 March 2002, Mark IV Industries and CDT Sign
08 April 2002, CDT Appoints Michael Unwin as Director
10 April 2002, Lord Sainsbury officially opens
02 May 2002, Cambridge Display Technology and Bayer Sign
14 May 2002, Sumitomo Chemical Places Equity Investment
14 June 2002, Seiko Epson Corporation and CDT Form Joint Venture
03 July 2002, TIME Magazine Names CDT one of Europe's hottest 50
08 July 2002, Light Emitting Polymer Pioneer Wins Award
15 July 2002, CDT Appoints Suk Bae Cha as Commercial Director
15 October 2002, Toppan Printing invests in CDT
28 October 2002, Cambridge Display Technology Adds Opsys OLED
30 October 2002, CDT Receive the MacRobert Gold Medal
11 November 2002, Cambridge Display Technology Named Technology

2003
12 March 2003, Cambridge Display Technology and Dupont
13 March 2003, CDT Receives First Order...
20 March 2003, UK Department of Trade and Industry Awards...
22 March 2003, Cambridge Display Technology Seeks Partner...
22 March 2003, Cambridge Display Technology Announce ULVAC...
27 March 2003, Cambridge Display Technology Achieves Milestone...
19 May 2003, Cambridge Display Technology Grants...
03 July 2003, CDT announces Reorganisation
22 July 2003, Luxell renews OLED Relationship
26 August 2003, Eastgate Technology is Granted License
21 November 2003, Descartes Prize

2004
February 2004, Sumitomo Co- Development of high efficiency LEP
17 March 2004, Philips announces Magic Mirror mobile phone model number 639
18 May 2004, Epson's 40-in OLED lights the way forward, Seiko Epson Corp.
25 May 2004, CDT and Kolon enter technology transfer program
25 May 2004, CDT & Covion report longer life blue
25 May 2004, 30K Blue Lifetime achieve with new polymer
27 July 2004, Descartes Prize - Japanese
27 July 2004, CDT Joins with Other Industry Leaders in New European Association
30 July 2004, Security Exchange Commission Form S-1, IPO filing for CDT
02 August 2004, Toppan Printing and CDTT in Joint PLED Printing Development Program
11 October 2004, CDT achieves 70k hour blue lifetime
12 October 2004, DELTA OPTOELECTRONICS and CDT sign major new agreement
07 December 2004, Seiko Epson will commercialize OLED TVs by 2007, PCWorld article 118852-page,1
16 December 2004, CDT Announces Flotation on NASDAQ

2005
28 January 2005, Merck KGaA sells electronic chemicals business to BASF
31 January 2005, High Efficiency P-OLED Materials Developed
07 February 2005, Joint Project to Investigate the use of Nanopartic
08 February 2005, Merck KGaA acquires Avecia's OLED and Polymer electronics units (www.merck.de)
14 February 2005, CDT WELCOMES SALE OF AVECIA UNITS TO MERCK KGaA
23 February 2005, TOPPAN PRINTING AND CAMBRIDGE DISPLAY TECHNOLOGY
04 March 2005, CDT AND AVI IN NEW BUSINESS COLLABORATION
04 March 2005, CAMBRIDGE DISPLAY TECHNOLOGY MAKES TWO SENIOR APPO
08 March 2005, Cambridge Display Technology Announces Financial R

14 March 2005, KEY NEW PATENTS ALLOWED IN PRINTABLE OLED DISPLAY
07 April 2005, CDT COLLABORATES WITH DELTA OPTOELECTRONICS TO REA
05 May 2005, CDT TO DEVELOP TOP EMISSION DISPLAY TECHNOLOGY
09 May 2005, Cambridge Display Technology Announces Financial R
19 May 2005, LARGE FORMAT INKJET PRINTERS SHIPPED TO MAJOR ASIA
23 May 2005, CDT ACHIEVES 100,000 HOUR BLUE POLYMER LIFETIME
24 May 2005, CDT CHIEF TECHNICAL OFFICER RECEIVES MAJOR PRIZE
24 May 2005, Joint Venture Announced in Polymer OLED material Supply
06 June 2005, CDT TO BRING ADVANCED TECHNOLOGY TO THE DISPLAYS M
10 August 2005, Cambridge Display Technology Announces Financial Report
17 August 2005, OTB Displays acquires Philips' PLED activity
18 August 2005, CDT STRENGTHENS OPERATIONS MANAGEMENT TEAM
07 September 2005, HIGHEST PERFORMANCE TEST EQUIPMENT NOW AVAILABLE
15 September 2005, CDT APPOINTS CHIEF FINANCIAL OFFICER
19 October 2005, JOINT VENTURE ANNOUNCED IN POLYMER OLED MATERIAL S
19 October 2005, CDT combines long lifetime with high efficiency
19 October 2005, Announcement of new Sumation joint venture
03 November 2005, MILESTONE IN OLED DISPLAYS DEMONSTRATED
09 November 2005, Cambridge Display Technology Announces Financial R
05 December 2005, CAMBRIDGE DISPLAY TECHNOLOGY APPOINTS NEW DIRECTOR
08 December 2005, SUMATION™ JOINT VENTURE COMPLETED
12 December 2005, CDT SEES NEW RECORD IN DISPLAY LIFETIME DEVELOPMEN
21 December 2005, CDT raises $17.5 million
01 December 2005, CDT NAMED ONE OF FASTEST GROWING UK TECHNOLOGY COM

2006
17 January 2006, CDT RECEIVES JEAN PIERRE NOBLANC AWARD FOR EXCELLE
13 February 2006, CDT ACQUIRES IMPORTANT NEW PATENT PORTFOLIO
13 March 2006, CDT SEES FURTHER RAPID PROGRESS IN POLYMER LIFETIME
13 March 2006, Cambridge Display Technology Announces Financial Report
16 March 2006, WORLD'S FIRST OLED PRINT HEAD USES POLYMER TECHNOLOLOGY
24 May 2006, CAMBRIDGE DISPLAY TECHNOLOGY AWARDED DTI GRANT FOR
04 May 2006, Cambridge Display Technology Announces Financial Report
06 June 2006, CAMBRIDGE DISPLAY TECHNOLOGY AND TOPPAN PRINTING T
06 June 2006, CAMBRIDGE DISPLAY TECHNOLOGY HELPS MERCK ENHANCE OLED
06 June 2006, CAMBRIDGE DISPLAY TECHNOLOGY AND LITREX LAUNCH DEVELOPMENT
06 June 2006, NATIONAL UNIVERSITY OF SINGAPORE STEPS UP ITS WORK
23 June 2006, CDT AND H.C. STARCK AGREE SUPPLY DEAL IN 'INK JETT
06 July 2006, CDT ANNOUNCES RESIGNATION OF COMPANY'S CHIEF FINANCE OFFICER
24 July 2006, CDT SUPPLIES INKJET PRINTING SOLUTION TO BRAZIL
14 August 2006, Cambridge Display Technology Announces Finance Report
18 August 2006, CDT COMMENCES RECRUITMENT OF AUDIT COMMITTEE CHAIR
24 August 2006, CDT Co-Operates in UK DTI-Supported Process Development
06 September 2006, CDT SEES RAPID PROGRESS IN BLUE POLYMER LIFETIME
20 September 2006, CDT Licensee MED Raises Funds and Commences Build
03 October 2006, FURTHER SIGNIFICANT PROGRESS IN POLYMER OLED LIFETIME
31 October 2006, CAMBRIDGE DISPLAY TECHNOLOGY ANNOUNCES MANAGEMENT
10 November 2006, Cambridge Display Technology Announces Financial Report
15 November 2006, CAMBRIDGE DISPLAY TECHNOLOGY ANNOUNCES MAJOR NEW
15 November 2006, CAMBRIDGE DISPLAY TECHNOLOGY ANNOUNCES ANOTHER 27 December 2006, SALES OF STOCK BY OFFICERS TO SETTLE TAX LIABILITIES LIFETIME

2007
01 January 2007, CDT Celebrates a Visit from the President of Sumitomo Chemical
04 January 2007, CDT Acquires Assets of P-OLED/OLED Display Driver

22 February 2007, CDT TO PRESENT AT UPCOMING INVESTOR CONFERENCES
28 February 2007, CAMBRIDGE DISPLAY TECHNOLOGY TO REPORT
01 March 2007, CDT Announces Financial Results for Fiscal 2006
12 March 2007, CDT Reports Verdict in Lease Dispute
26 March 2007, CDT and Sumation Announce Strong Lifetime Improve
26 March 2007, Ian Chao Promoted to Vice President of Commercial
07 April 2007, Sony's 11-inch OLED HDTV Press Release, Sony.net Press_Archive/200409/04-048E
12 April 2007, CAMBRIDGE DISPLAY TECHNOLOGY AWARDED DTI GRANT FOR WORK ON FLUID MODELLING
18 April 2007, Hilary Charles Joins CDT as General Counsel
19 April 2007, CDT's P-OLED Technology Featured
27 April 2007, CDT to Present at the AeA Micro Cap Financial Conf
02 May 2007, CDT LICENSEE MICROEMISSIVE DISPLAYS TO BEGIN INITI
10 May 2007, CDT to Report First Quarter 2007 Results on May 15
15 May 2007, CDT Announces First Quarter 2007 Financial Results
21 May 2007, CDT and SUMATION Announce Improved Performance Cha
22 May 2007, CDT TO SHOWCASE HIGH RESOLUTION P-OLED BASED QVGA
16 July 2007, CDT AWARDED ADDITIONAL DEPARTMENT OF TRADE AND IND
31 July 2007, Sumitomo Chemical Company to Acquire CDT
15 August 2007, CDT Announces Second Quarter 2007
27 August 2007, CDT and Sumation Announce Further Improved Perform
19 September 2007, Sumitomo Chemical Company Completes Acquisition of CDT
03 October 2007, CDT CELEBRATES A VISIT FROM THE PRESIDENT OF SUMITOMO

3. Seybold Publications (1998) *Gretag Imaging to acquire Raster Graphics* http://www.seyboldreports.com/SRPS/subs/2804/html/news6.htm
4. Fyfe, David, (2006) SID Investors Conference Presentation
5. 2007, July 31 Sumitomo Chemical to Acquire Cambridge Display Technology Inc.
6. 2004-2007 NASDAQ Historical stock prices for OLED (Cambridge Display Technologies
7. Fyfe, David (Presentation) *Invention to Commercialization*, Lux Executive Summit, October 2007
8. McLaughlin Consulting Group & Insight Media (2004) *The world of CDT and its partners: Micro Emissive Displays*, Case Study
9. Burroughes, Jeremy, Content obtained through extensive interviews and meetings
10. Fyfe, David, Content obtained through extensive interviews and meetings

Chapter 12 Summary

1. Ramon Marimon (2003) "University Research Business and Local Development", Luiss Guido Carli University
2. Post-It® is a registered trademark of 3M
3. iPOD® is a registered trademark of Apple Inc.
4. iTunes® is a registered trademark of Apple Inc.

Index

Notes: